the
UNIVERSITY
of
GREENWICH

PERGAMON INTERNATIONAL LIBRARY
of Science, Technology, Engineering and Social Studies

*The 1000-volume original paperback library in aid of education,
industrial training and the enjoyment of leisure*

Publisher: Robert Maxwell, M.C.

Sun Power

AN INTRODUCTION TO THE APPLICATIONS
OF SOLAR ENERGY

SECOND EDITION

THE PERGAMON TEXTBOOK
INSPECTION COPY SERVICE

An inspection copy of any book published in the Pergamon International Library will
gladly be sent to academic staff without obligation for their consideration for course
adoption or recommendation. Copies may be retained for a period of 60 days from
receipt and returned if not suitable. When a particular title is adopted or recommended
for adoption for class use and the recommendation results in a sale of 12 or more copies,
the inspection copy may be retained with our compliments. The Publishers will be
pleased to receive suggestions for revised editions and new titles to be published in this
important International Library.

Other Pergamon Titles of Interest

Pergamon Related Journals *(free specimen copy gladly sent on request)*

Sun Power

AN INTRODUCTION TO THE APPLICATIONS OF SOLAR ENERGY

SECOND EDITION

By

J. C. McVEIGH

M.A., M.Sc., Ph.D., C. Eng., F.I. Mech.E., F. Inst.E.,
M.I. Prod.E., M.C.I.B.S.

Head of Energy Studies,
Brighton Polytechnic

PERGAMON PRESS

OXFORD · NEW YORK · TORONTO · SYDNEY · PARIS · FRANKFURT

U.K.	Pergamon Press Ltd., Headington Hill Hall, Oxford OX3 0BW, England
U.S.A.	Pergamon Press Inc., Maxwell House, Fairview Park, Elmsford, New York 10523, U.S.A.
CANADA	Pergamon Press Canada Ltd., Suite 104, 150 Consumers Rd., Willowdale, Ontario M2J 1P9, Canada
AUSTRALIA	Pergamon Press (Aust.) Pty. Ltd., P.O. Box 544, Potts Point, N.S.W. 2011, Australia
FRANCE	Pergamon Press SARL, 24 rue des Ecoles, 75240 Paris, Cedex 05, France
FEDERAL REPUBLIC OF GERMANY	Pergamon Press GmbH, Hammerweg 6, D-6242 Kronberg-Taunus, Federal Republic of Germany

First edition 1977
Second edition 1983

British Library Cataloguing in Publication Data
McVeigh, J.C.
Sun power.—2nd ed.—(Pergamon international library)
1. Solar energy
I. Title
621.47   TJ810
ISBN 0-08-026148-5 (Hardcover)
ISBN 0-08-026147-7 (Flexicover)

In order to make this volume available as economically and as rapidly as possible the typescript has been reproduced in its original form. This method unfortunately has its typographical limitations but it is hoped that they in no way distract the reader.

Printed in Great Britain by A. Wheaton & Co. Ltd., Exeter

This book is dedicated to those who are striving to conserve rather than consume, to develop a simpler non-violent lifestyle and to obtain power without pollution through the use of renewable natural energy resources.

PREFACE TO THE SECOND EDITION

The past six years have seen many new developments in the applications of solar energy and the completion of projects which were in their early stages when the first edition of Sun Power was published. Among these could be listed the three hundredfold increase in the maximum size of the photovoltaic array, the hundredfold increase in the maximum size of the solar power tower and the fiftyfold increase in the power output from a solar pond. My original aim was to provide a general introduction to the very wide range of these solar applications. The support I have received from many friends and colleagues throughout the world has encouraged me to follow the same pattern in this extensively revised edition.

Some of the material originally in the first edition has either been shortened or omitted to make way for more recent developments. For example in the field of solar collectors several companies have developed second generation collectors with greatly improved performances in the higher temperature ranges with lower radiation conditions. Some of these trends are discussed in Chapter 3. With the considerable emphasis in passive space heating systems seen during the past six years they clearly merit a separate chapter. Social and legal issues have been introduced in the chapter which includes methods of economic analysis while the chapter on photovoltaic cells, biological conversion systems and photochemistry has been considerably extended. I have continued to keep the theoretical treatment to a minimum, an approach which enables a broad appreciation of the subject to be obtained in a fairly short period. For those who wish to pursue their studies further, a selection from some of the more rigorous textbooks is reviewed briefly in an appendix. There are now approximately six hundred references in this edition with nearly half being from work completed since 1976. These have been largely drawn from widely available journals and conference proceedings and should provide immediate access to current research areas — a feature of the first edition. As the literature has expanded so rapidly there are inevitably gaps in the coverage. For example, while translations from the Soviet literature are available (Gelioteckhnika, Appendix 3), comparatively few Western Institutes can afford a further journal subscription for a small minority of research workers. This is unfortunate as some of the Soviet work, especially in passive heating techniques applied to greenhouses, could have important applications in other countries.

In preparing this edition I have had helpful discussions with many people, especially Raymond Tomkins (Imperial College) who corrected an error in my original economic analysis, Royston Summers and Doug Kelbaugh who discussed their solar houses and provided illustrations (5.2 and 5.4 respectively), Bernard McNelis who also provided illustrations 6.5 and 8.2, the community at Twind who provided illustration 9.2, Dr Raja Hajjar (University of Beirut) who provided information from the Arab Countries, Judith Stammers, executive director of the Solar Trade Association (UK) who also gave permission to include an extract from their Code of Practice, June Morton, administrator of UK-ISES, Dr Leslie Jesch (University of Birmingham), Owen Lewis (University College Dublin), Simos Yannas (Architectural Association Graduate School) and Colonel Omer Sharfi and Carlo Maselkowski (formerly post-graduate students at Brighton Polytechnic). In the final stages Laura Mangan provided valuable typing and editorial assistance and it was a pleasure to work again with Jim Gilgunn-Jones and his colleagues of Pergamon Press.

I would like to thank Ewbank and Partners of Brighton for their encouragement and support for my solar energy activities and I acknowledge the continued support from the Director and Council of the Brighton Polytechnic.

There are still many problems awaiting solutions. At the Solar World Forum held in Brighton during August 1981 Professor Jack Duffy identified several, including the storage of energy in the temperature range above 100°C in liquid systems and looking more deeply into aspects of control in the simulation and modelling of solar systems. In parallel with technical solutions we must also find ways of bringing solar education within reach of everyone at the appropriate level. Perhaps the most important lesson which those of us who were privileged to be present at Brighton in 1981 learnt was that solar energy is a force for peace.

Brighton, Sussex   Cleland McVeigh
November, 1982

PREFACE TO THE FIRST EDITION

We are living at a time when there is a greater awareness of the energy problems facing the world than at any other period in history. It is now widely accepted that the growth in energy consumption which has been experienced for many years cannot continue indefinitely as there is a limit to our reserves of fossil fuel. Solar energy is by far the most attractive alternative energy source for the future, as apart from its non-polluting qualities, the amount of energy which is available for conversion is several orders of magnitude greater than all present world requirements.

It is exciting and challenging to realize that we can all share in this inexhaustible energy source. In the long term some of the projects described in my book as being in their early stages of development will be providing a substantial part of the energy requirements of many countries. There are already several thousand solar water heating systems in the United Kingdom and each system is saving energy and helping the country to become more self-sufficient in energy resources. The amount of money that a solar system saves annually will always keep pace with increasing fuel costs. By providing a general introduction to the very wide range of solar applications currently being pursued throughout the world I hope that many others will be encouraged to install their own solar systems, which will indirectly save many other resources such as water and building materials.

While a basic scientific knowledge would be an advantage for a complete understanding of each topic, the theoretical treatment has been reduced to a minimum and many of the applications are illustrated with original photographs, diagrams or sketches. The main emphasis is on thermal applications such as water heating, space heating and cooling and small scale power, which should make an immediate impact in many countries during the next five years. Practical construction details of several solar heaters and systems are included, as well as methods of assessing the economics of solar systems.

The applications are arranged in chapters which are intended to be complete in themselves and can be studied separately. This approach has meant that a few topics, such as the practical details of solar heating systems, appear in two chapters although the treatment is different. Architects and engineers will find that the reference sections, which contain over 350 references, are sufficiently comprehensive for immediate access to current

research areas. Some of the references are to papers published originally
as abstracts in the 1975 International Solar Energy Congress and
subsequently in the journal Solar Energy in 1976. There is a considerable
variation in the units used in the source material and generally these have
been converted to SI units.

During the preparation of this book I have been helped either directly or
indirectly by many people and it is not possible to mention them all
individually. My original interests were in the thermal applications of
solar energy and I am particularly grateful to those who have enabled me to
have closer contact with different interests and new applications, especially
my colleagues on the Committee of the UK Section of the International Solar
Energy Society. Professor John Page, the first Chairman of the UK Section
of ISES, has kindly allowed me to use some of his derived radiation data,
and I have benefited from frequent discussions with him on the wider issues.
It was a rewarding experience to form part of his team which prepared the
1976 report on Solar Energy, a UK Assessment, and some of the material which
appears in Chapters 3, 4 and 5 is based on work which I prepared for that
publication. I have also had helpful discussions with Professor David Hall
(Biology), Professor Peter Landsberg (Economics), Dr. Mary Archer
(Photochemistry) and Michael Blee, Edward Curtis, Dominic Michaelis and Don
Wilson (Architecture and Solar Space Heating). Philip Baxendale suggested
ideas for my early experiments on domestic water heaters in the 1960s and
technical assistance has been provided by Roy Davis (Woolwich Polytechnic)
and George Moore and Dave Burton (Brighton Polytechnic). Dr. Frances
Heywood very kindly allowed me to have access to the collected papers of the
late Professor Harold Heywood.

My thanks are also due to many friends and colleagues at the Brighton
Polytechnic. Advice on collector materials and corrosion was given by John
Lane. Dr. Alastair McCartney prepared Figure 2.2 and advised on solar
radiation data. The eight full-page illustrations were drawn by Leonore
Duff. Garry Hibbert prepared many of the photographs. Chris Wimlett traced
the line diagrams and Nancy Holmes typed a preliminary draft. The final
typescript which appears in this edition was prepared by Celia Rhodes. I
also acknowledge the continued support for my solar energy work from the
Director and Council of the Brighton Polytechnic.

Some of the meteorological data in Chapter 2 and the whole of Appendix 2 is
reproduced with the permission of the Controller of Her Majesty's Stationery
Office. I acknowledge with thanks the permission given by the following
organisations and individuals to use their illustrations: The Copper
Development Association (4.7), Edward Curtis and Homes and Gardens (4.8),
The Electricity Council Research Centre, Capenhurst (4.10), the Philips group
(4.15), Ferranti Ltd. (7.2) and Geoffrey Pontin and the Wind Energy Supply
Co. Ltd. (8.3). Senior Platecoil Ltd. have allowed me to quote from their
brochure in Chapter 9. The Editor of the Building Services Engineer
(formerly the JIHVE) has allowed me to quote directly from material published
in the June 1971 edition of the JIHVE.

Finally, it was my great privilege to serve as a Departmental Head under the
late Professor Harold Heywood at the Woolwich Polytechnic some ten years ago.
He inspired many workers through the excellence of his research and his
helpful suggestions and enthusiasm laid the foundations for my own interests
in solar energy.

Brighton, Sussex   Cleland McVeigh
September, 1976

CONTENTS

CHAPTER 1

INTRODUCTION AND HISTORICAL BACKGROUND

Man has appreciated for thousands of years that life and energy flow from the sun. Socrates (470-399 B.C.) is believed to have been the earliest philosopher to describe some of the fundamental principles governing the use of solar energy in applications to buildings as the following passage from Xenophon's Memorablia indicates:

In houses with a southerly aspect, the rays of the sun penetrate into the porticos during the winter, but in the summer the sun's path is directly over our heads and above the roofs, so that there is shade. If, therefore, this is the best arrangement, we should build the south side higher, to catch the winter sun, and the north side lower to exclude the cold winds . . .

Another early application was the alleged attack by Archimedes upon the Roman fleet at Syracuse in the year 214 B.C. He is supposed to have constructed a number of highly polished focusing metal mirrors which were arranged along the shore, so that the reflected rays of the sun were concentrated upon the hulls and rigging of the Roman ships which lay in the harbour or sailed inshore. Some of them were set on fire and this caused the Roman fleet to scatter. Most of the early applications related to various focusing systems, such as mirrors or lenses. Anthemius de Tralles, a celebrated architect in the sixth century, left among his papers four treatises on burning mirrors. One of these treatises is entitled "How to construct a machine capable of setting an object on fire at a distance by means of solar rays". An English monk, Roger Bacon, also worked on burning mirrors, towards the end of the thirteenth century. The first solar operated water pump was invented by Salamon de Caus (1576-1626) a French engineer who described this machine in 1615. The French philosopher Buffon carried out various experiments in 1747 which demonstrated the practicability of the attack at Syracuse. He built a large framework on which he hung pieces of silvered glass which could reflect to a single focal point. He then varied the number of mirrors and the focal point so that ultimately at a distance of 77 m using 154 mirrors he burned chips of wood covered with charcoal and sulphur. Subsequently he constructed a parabolic mirror with a diameter of 1.17 m but all his experiments were regarded as, at best, scientific curiosities by his contemporaries. An early example of a solar cooker is recorded by De Saussure, a Swiss philosopher (1740-1799) in letters to *Buffon* and *La Journal de Paris*. These describe how he constructed a set

1

of concentric glass chambers and cooked soup in the centre. A French
physicist is credited with a similar discovery at about the same time.
Bernard Foret Belidor (1697-1761) also invented a form of solar operated
pump, or continual fountain, which is shown in Fig. 1.1.

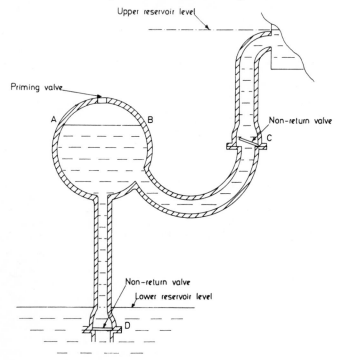

Fig. 1.1. Belidor's solar pump. The pump is primed by
           filling the spherical dome to the level AB.
           During the day solar radiation heats the dome,
           causing the air to expand and force the water
           through the non-return valve C to reach the
           upper reservoir. On cooling, either artifi-
           cially or at night, the internal air pressure
           falls below atmospheric, drawing water into
           the pump from the lower reservoir through the
           non-return valve D.

Experiments to determine the intensity of the sun's radiation - the solar
constant - were first carried out at the beginning of the nineteenth century
by Sir John Herschel, who invented an actinometer "an instrument for
measuring the intensity of heat in the sun's rays", and, quite independently,
by the French scientist Pouillet. Both used the same principle - exposing a
known quantity of water to solar radiation and measuring the temperature
rise which occurred over a given period of time. Herschel's actinometer was
a stationary open vessel while Pouillet's instrument was a closed movable
vessel - the pyrheliometer. Their calculations included allowances for
atmospheric absorption and scattering or attenuation. Pouillet's instrument
and experimental methods were severely criticized by John Ericsson (1), who
commented that computations based on latitude, date and exact time were too
complex and tedious for investigations in which the principal element, the

depth of the atmosphere, was continually changing. Ericsson is better known for his caloric engine and hot air cycle, but he was also a brilliant naval architect before he turned to experiments with solar energy. His solar calorimeter was fixed to a "vibrating table applied within a revolving observatory, supported on horizontal journals and provided with a declination movement and a graduated arc". On 7 March 1871, he stated that "the dynamic energy developed on one square foot of surface at the boundary of the atmosphere is 7.11 units (Btu) per minute". This is equivalent to 1.332 kW/km^2 - a figure which is quite remarkable as it is within 3% of the latest accepted estimates of the value of the solar constant (1.373 kW/m^2).

The earliest recorded instance of a patent relating to solar energy (2) dates from 1854, when Antoine Poncon was granted a patent in London for ". . . using the sun's rays to create a vacuum in a suitable vessel, elevated at the height of a column of water, which, in the above vacuum, is kept in equilibrium by the pressure of the atmosphere. Such vacuum being formed, I fill it with water acted upon by the external pressure of the atmosphere, and thus obtain the heat of water which may be obtained as a motive power." Since then, nothing has been discovered about this invention and there are no records of its construction. Various other British patents occur over the next few years, but it is highly unlikely that the inventions which were claimed had ever been constructed. In contrast to these theoretical ideas, a French Professor, August Mouchot, had apparently constructed a parabolic focusing mirror which he used to drive a small steam engine in 1860 and he received a patent from the French Government in 1861. Subsequently he exhibited a "solar pumping-engine" in Paris in 1866 and also experimented with solar cookers. He wrote the first book ever published on solar energy, "La Chaleur Solaire et Ses Applications Industrielles" (3) in 1869, and on 29 September 1878 successfully demonstrated the first solar-operated refrigerator, producing a block of ice at the Paris Exposition.

Although Ericsson claimed in 1868 that he had constructed the first solar engines, it would seem that Mouchot was several years ahead of him. Ericsson was certainly the first inventor of a solar engine operating on an air cycle and this engine is reported to have worked in New York in 1872, "at a steady rate of 420 rev/min when the sun was in its zenith and the sky clear." Designed primarily for low powered applications, Ericsson noted that it was better to use steam in the engine when high powers were required.

With his considerable knowledge of solar radiation and his earlier experience in naval architecture and mechanical engineering, it is not surprising that Ericsson was also concerned about the energy crisis in 1876. He predicted that the coal fields would eventually be exhausted and that this would cause great changes in international relations in favour of countries with continuous sun power. He carried out a thoeretical analysis on the use of solar engines on a strip of land about 8000 miles long and one mile wide (about 13 000 km by 1.6 km) and commented:

> Upper Egypt for instance, will, in the course of time, derive signal advantage and attain a high political position on account of her perpetual sunshine, and the consequent command of unlimited motive force. The time will come when Europe must stop her mills for the want of coal. Upper Egypt, then, with her never-ceasing sunpower, will invite the European manufacturer to remove his machinery and erect his mills on the firm ground along the sides of the alluvial plain of the Nile where sufficient power can be obtained to enable him to run more spindles than 100 such cities as Manchester.

Economic reasons also led to the development of the first, and, for many years, the world's largest solar distillation system in Las Salinas, about 110 km inland from the coast of Chile. The local water, which contained about 14% salts, was quite unsuitable for use in steam boilers and there was also the problem of supplying large quantities of drinking water for animals and men. A complete specification of the system, which was designed by Charles Wilson in 1872, was given by Harding (4). It consisted of 64 frames, each 60.96 m long by 1.22 m broad, giving a total area of 4756 m^2 of glass. A particular feature of the plant was an early application of self-sufficiency as the salty water was pumped from the local wells by a windmill into a storage tank at the highest end of the plant. Initially some 19 000 litres of fresh water could be produced daily at a cost of approximately one quarter that of a conventional coal-fired boiler system, but after a railway line had been built, the demand for the water decreased and the whole system was allowed to disintegrate.

The earliest United States patent relating to some form of focusing device was claimed by Charles Pope, a parson, in 1875. Pope was fascinated by the wide variety of solar applications which were opening up at the time and eventually produced and published in Boston the first book written in English on solar energy in 1903 - "Solar Heat - its practical applications" (2). The first major solar patent in the United States was issued on 20 March 1877 to John S. Hittell and George W. Deitzler of San Francisco. Their patent describes "a concave mirror by which they throw focalized heat on a mass of iron or other suitable material as a reservoir of heat . . . . letting the cold air pass in and then pass out again after the sun has heated it, applying it then to ordinary hot air machinery" (the Ericsson cycle). Deitzler took out a second patent on 19 May 1882 for a reflecting mirror and was a founder director of the Solar Heat Power Company of California in 1883.

India was another country in which early work was carried out. Mr. W. Adams, an English resident of Bombay, invented a solar cooker consisting of a conical reflector 0.711 m in diameter made of wood and lined with common silvered cheap glass. "The rations of seven soldiers, consisting of meat and vegetables, was thoroughly cooked by it in 2 hours in January, the coolest month of the year in Bombay" (5).

Abel Pifre continued Mouchot's work in France and, on 6 August 1882, used a mirror 3.5 m in diameter to power a small vertical steam engine which worked a Marinoni printing press in Paris. Although there was some cloud in the sky, an average of 500 copies per hour of "Soleil-Journal", a journal specially composed for the occasion, were printed between 1 p.m. and 5 p.m.

One of the earliest space heating applications was claimed in 1882 by Professor E. S. Morse of Salim, Massachusetts in an invention for "Utilizing the sun's rays in warming houses" (6). It consisted of a surface of blackened slate under glass, fixed to the sunny side of a house, with vents in the wall arranged that the cold air in a room was let out at the bottom of the slate and forced in again at the top by the ascending heated column between the slate and the glass. This method was used to heat Professor Morse's own house in fine weather. Also at about the same period the first use of the flat plate collector is reported (7), but in an application to a water pumping system.

The first commercial solar water heaters in the world appeared in the United States shortly afterwards, manufactured by Clarence M. Kemp from Baltimore, Maryland, who patented the Climax Solar-Water Heater in 1891 (8). This new

industry developed steadily, particularly in California where several
thousand systems had been installed by 1900. At the same time the size of
the solar engine industry began to increase greatly. An anonymous group of
"Boston capitalists" developed several engines and, in 1901, their most
successful engine was described in various publications (9, 10). The engine
was erected at an ostrich farm in South Pasadena, California and consisted
of a reflector, 10.2 m in diameter at the top and 4.57 m in diameter at the
bottom, with an inner surface consisting of 1788 mirrors, approximately
99 mm x 600 mm, arranged to focus on a suspended boiler. The reflector stood
on an equatorial mounting, with a north-south axis and tracked the sun, from
east to west, by means of a clockwork mechanism. There is some doubt about
its actual performance. Although an output of 15 hp has claimed, the actual
recorded daily average when pumping water was only about 4 hp.

The other groups who were involved in large-scale engine testing at that time
were the Shuman Engine Syndicate Ltd., and the Sun Power Company (Eastern
Hemisphere) Ltd. Their developments are very fully recorded by one of their
consultants, A. S. E. Ackermann (11) in 1914. Mr. Frank Shuman's 1907
prototype consisted of a number of parallel horizontal black pipes containing
ether, placed on the ground in a shallow box about 6 m x 18 m x 0.45 m deep,
containing water with a layer of melted paraffin wax on it under a glass
cover. The ether boiled and produced vapour at a sufficient pressure to
drive a small vertical reciprocating engine. The exhausted ether was
condensed and recycled. The second design was built at Tacony, Philadelphia
in 1910 and was quite different in principle, using water only. A flat plate
boiler was constructed from two thin copper sheets each 1.83 m long by
0.76 m wide with a narrow gap between them for the water. Cold water was
admitted at the lower edge at one corner and a steam pipe was attached to
the upper edge of the upper corner. The boiler was placed in a double
glazed, insulated wooden box which was mounted on an east-west axis. No
attempt was made to track the sun, but the inclination of the box was
adjusted weekly so that the glass top was perpendicular to the sun's noon
position. The system was successful in raising steam. The following year
a full-size system was built, with 956.5 m of collector area and the use of
plain glass mirrors to achieve a concentration ratio of 2:1. No satisfactory
method of measuring the actual output from this plant was available, but a
maximum estimated 26.8 bhp was obtained by matching the steam conditions with
the earlier test results.

The group then invited Professor C. V. Boys to join them in which became the
most spectacular solar engine development of the time - the Shuman-Boys Sun-
Heat Absorber at Meadi in Egypt. Professor Boys improved the Tacony design
by introducing an automatic tracking system. The absorber consisted of five
large parabolic mirror sections, each 62.5 m long and 4.1 m wide between the
edges of the mirrors, giving a total collecting area of 1277 m^2. Each
mirror was built up from various sizes of flat glass coated with shellac.
They were carried on a light framework of painted steel and each of the
sections was driven by a system of tubular shafting which caused the
parabolic mirrors to rotate. They were placed with the major axes north-
south. Each morning they were heeled over to the east and then they moved
automatically and slowly from that position to the west, tracking the sun.
An extensive series of tests were carried out in 1913, with a maximum
recorded pumping horsepower of only 19.1 hp. Ackerman commented that this
was an exceedingly bad result which he attributed to the engine and pumping
side of the plant. Calculations which he based on the performance of another
steam engine which he tested in England show that the steam conditions at
Meadi could have developed 55.5 bhp.

At this point it can be seen that although only a restricted range of
engineering materials was available, the basic principles of many practical
applications of solar energy had been understood and that much of the work
had required very considerable technical expertise. However, the era of
cheap alternative fuel resources had commenced and the next two decades saw
a period of comparatively little interest in solar energy as the development
of first oil and then gas had priority. For example, the solar water
heating industry, which was well established in parts of California in the
mid-1920s, had practically disappeared by the end of the 1930s. Fortunately
some dedicated workers, such as Dr. G. C. Abbot in the United States
continued to carry out original research, but it was not until the early
1940s that a resurgence of interest in the utilization of solar energy took
place. This was encouraged by the Godfrey L. Cabot bequest to the
Massachusetts Institute of Technology to promote solar energy research and
from this renaissance the work spread to research teams in many parts of the
United States and to other countries of the world. The first major symposium
on wind and solar energy was held in New Delhi in October 1954 (12) and the
need to establish closer links between the various countries led to the
formation of the Association for Applied Solar Energy, now the International
Solar Energy Society (ISES). This has as its purpose the encouragement of
basic and applied solar energy research, the fostering of the science and
technology relating to applications of solar energy and the compilation and
dissemination of information relating to all aspects of solar energy. The
New Delhi Symposium was followed in November 1955 by two conferences in
Arizona, the first, on basic research, was held at the University of Arizona
(13) and the second was a world symposium at Phoenix (14), where a large
variety of solar equipment was displayed, including radiation measuring
instruments, water and air heaters, cookers, models of various solar houses,
high temperature furnaces, water stills, photovoltaic converters and several
different types of engine, the largest developing about 2.5 hp.

Several other conferences were held during the next 15 years. In 1961 the
United Nations held a symposium in Rome on new sources of energy (15) and
there was also an international seminar in Greece (16). The 1970 ISES
Conference in Melbourne was the last of the pre energy-crisis era.

Two major reports were published shortly before the UNESCO Conference, "The
sun in the service of mankind" in Paris, July 1973, the first from the
United States (17) and the second from Australia (18). Both reports high-
lighted the benefits which their respective countries could obtain from solar
energy applications. In July 1975 the largest conference on solar energy
held up to that point took place at the University of California, Los Angeles
(UCLA), with 265 technical papers presented, over 60 different commercial
exhibits and a total attendance estimated at more than 1700. Among many
countries who published their own 'solar reports' at this time were Ireland
(19) and the UK (20).

Interest in solar energy research and development started to spread very
rapidly and an impression of the extent of world activity in 1976 can be
obtained from Table 1.1, which was largely compiled from surveys published in
the UK (21, 22), USA (23) and the EEC (24). From the various sources quoted
in these surveys it was clear that this table underestimated the position,
as it proved difficult in some cases to obtain a response to the requests for
information. The classifications in Table 1.1 were necessarily rather broad
and included economic or theoretical studies.

Table 1.1.   Some solar energy research and applications in 1976

	1	2	3	4	5	6	7	8	9	10
Argentina	X	X	X		X	X		X		X
Australia	X	X	X		X	X		X	X	X
Austria	X	X		X				X	X	X
Belgium	X	X						X		X
Canada	X	X	X	X		X	X	X	X	X
Costa Rica	X	X	X	X			X			X
Denmark	X	X	X				X	X		X
Ecuador	X	X								X
Finland								X		
France	X	X	X	X		X	X	X	X	X
Greece	X	X		X	X					
India	X	X	X		X	X	X	X	X	X
Iran	X	X	X		X	X		X	X	X
Iraq	X				X					X
Ireland	X	X				X	X	X	X	X
Israel	X	X	X				X			
Italy	X	X	X	X	X			X	X	X
Jamaica	X		X			X				
Japan	X	X	X	X			X	X		X
Jordan	X	X	X		X	X				X
Korea	X	X								
Kuwait	X							X		
Netherlands	X	X					X	X		X
New Zealand	X		X							X
Nigeria			X							X
OAS*	X		X					X		X
Pakistan	X		X		X	X		X		
Papua New Guinea			X							
Saudi Arabia	X	X			X					X
South Africa	X	X	X				X			X
Spain	X			X	X			X	X	
Sri Lanka	X		X			X				X
Sweden	X	X					X			X
Turkey	X	X	X		X	X		X	X	X
UK	X	X	X			X	X	X	X	X
USA	X	X	X	X	X	X	X	X	X	X
USSR	X	X	X	X	X	X	X	X	X	X
West Germany	X	X	X		X	X	X	X		X
West Indies	X		X		X	X	X		X	X

Key to column headings
1. Water heating, including domestic and industrial processes.
2. Space heating, solar houses, applications in architecture.
3. Smaller thermal applications, pumping, engines, refrigeration and cooling.
4. Large-scale thermal applications, including focusing devices and furnaces.
5. Desalination and distillation.
6. Agricultural, including crop drying.
7. Wind power.
8. Photovoltaic and photochemical applications.
9. Photobiology, bioconversion.
10. Radiation studies.

*OAS - The Organization of American States.

The next ISES Congress, held in New Delhi in 1978 had even more papers, 450, and national programmes were reported from five countries – Malawi, Mexico, the Philippines, Thailand and Vietnam – not included in Table 1.1. A significant item was the report from the United Nations on the wide ranging work of their various organizations (25, 26). The decade ended with the ISES Silver Jubilee Congress in Atlanta (27), where the number of participants exceeded 2000 for the first time.

By the end of 1980 overviews of the prospects for solar energy were appearing regularly in the literature – see, for example, references (28) and (29).

National sections of ISES were spreading information about research development and demonstration projects. Typical examples were conferences on European Housing (30) and Codes of Practice and Test Procedures (31) organized by the UK section.

Solar industries were well established in most European countries. Typical examples included the UK exporting solar-related goods and services to the value of over £5 million annually; the German solar industries association, BSE, having a membership of around 40 large companies; a Belgian solar trade association with about 30 companies; Italy reporting some 60 commercial companies and Spain about 50 (32).

In the United States, the massive government support for solar energy programmes over the past ten years had developed many new industries and educational programmes. Publications on solar energy had expanded at an exponential rate, with well over 100 journals containing solar-related topics.

The Arab countries are attaching great importance to the effective exploitation of solar energy as shown in Table 1.2, which was prepared from the United Nations report in 1981 (33). Of the 21 countries listed in the table only five were known to be actively involved a few years earlier, as shown in Table 1.1.

The financial resources available from one country to another sometimes differ by a factor of a thousand or more, but a feature common to nearly every programme is an understanding that it does not take a substantial research and development programme to make a worthwhile contribution. The world's largest programme started in the United States with a modest $1.2 million. Its growth to $300 million by 1977 is illustrated in Fig. 1.2. By the financial year 1980/1, estimates for the annual expenditure were in the order of $1500 million before cuts imposed by the Reagan administration reduced solar funding for the first time. Estimates for 1981/2 were still substantial, at over $200 million, but no longer so far ahead of the rest of the world. France had spent about $77 million in 1981, Japan $63 million and both Germany and Italy in the order of $40 million. The rapidly increasing budget of the Arab countries was estimated to be over $25 million in 1980.

Table 1.2. Summary of existing and recent solar energy
research and applications in Arab countries

	1	2	3	4	5	6	7	8	9	10	11	12
Algeria	X	X	X	X		X		X			X	
Bahrain	X		X				X					X
Djibouti								X				
Egypt	X	X	X	X	X	X		X	X	X	X	X
Iraq	X	X			X	X	X					X
Jordan	X	X	X	X	X	X	X				X	X
Kuwait	X	X		X	X		X		X	X	X	X
Lebanon	X				X	X				X	X	X
Libya[1]	X				X	X						X
Mauritania								X		X		
Morocco										X	X	X
Oman	X									X		
Qatar											X	
Saudi Arabia	X	X	X	X	X			X	X	X	X	X
Somalia					X							
Sudan	X		X		X	X		X			X	X
Syria[2]	X	X								X	X	
Tunisia	X				X			X	X	X	X	X
UAE[3]						X	X	X				
North Yemen[4]	X	X					X	X			X	X
South Yemen[5]									X	X		

The country names listed above were used in the original United Nations
Economic Commission for Western Asia reports. Official United Nations
designations are given below:
[1]Libyan Arab Jamahiriya.
[2]Syrian Arab Republic.
[3]United Arab Emirates.
[4]Yemen.
[5]Democratic Yemen.

Key to column headings
 1. Water heating and/or flat plate collectors.
 2. Space heating, solar houses, applications in architecture.
 3. Cooling and refrigeration.
 4. Thermal conversion, including focusing devices and furnaces.
 5. Desalination and distillation.
 6. Crop drying.
 7. Agricultural greenhouses.
 8. Water pumping.
 9. Storage.
10. Wind power.
11. Photovoltaic conversion.
12. Radiation studies.

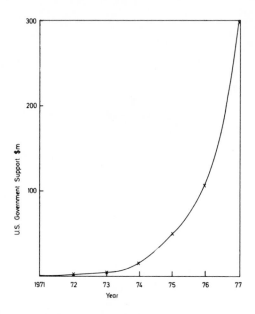

Fig. 1.2.

Conventional fossil fuel reserves can only last perhaps a hundred years at
the most and there are very considerable technical and environmental
reservations about nuclear energy. Solar energy, which is non-polluting and
inexhaustible, is already economically viable for certain applications in
almost every country in the world. Political decisions to invest in solar
research, development and demonstration programmes have already been taken
in many countries. For those who are in a position to influence national
energy policies towards an increasing use of solar energy there is only one
message - time is short.

REFERENCES

(1) Ericsson, J., *Contributions to the Centennial Exhibition*. John Ross,
     New York, 1876.

(2) Pope, C. H., *Solar Heat - its practical applications*. Boston,
     Massachusetts, 1903.

(3) Mouchot, August, *La Chaleur Solaire et Ses Applications Industrielles*.
     Gauthier-Villars, Paris, 1869.

(4) Harding, J., Apparatus for Solar Distillation. Paper No. 1933,
     Selected Papers, *Institution of Civil Engineers* 73, 1908.

(5) Adams, W., Cooking by Solar Heat, *Scientific American*, 19 June, 1878.

(6) *Scientific American*, 13 May, 1882.

(7)   The utilization of solar heat for the elevation of water, *Scientific American*, 3 October, 1885.

(8)   Butti, K. and Perlin, J., *A Golden Thread*, Cheshire Books, Palo Alto and Van Nostrand Reinhold, New York, 1980.

(9)   A solar motor, *The Engineering Times* 5, 186-187, April 1901.

(10) Thurston, R. H., Utilizing the sun's energy, *Cassiers Magazine*, New York, August 1901.

(11) Ackerman, A. S. E., The utilisation of solar energy, *Trans. Soc. of Engs*. 81-165, 1914.

(12) Wind Power and Solar, Proc. New Delhi Symposium 1954, Paris, UNESCO, 1956.

(13) Trans Conf. on the Use of Solar Energy - the Scientific Basis, Tucson, Arizona, November 1955. University of Arizona Press, 1958.

(14) Proc. World Symposium on Applied Solar Energy, Phoenix, Arizona, November 1955, Stanford Research Institute, Menlo Park, California, 1956.

(15) UN Conf. on New Sources of Energy, Rome 1961. Proceedings, 4-6, New York, United Nations, 1964.

(16) Spanides, A. G. and Hatzikakidis, A. D., eds, Solar and Aeolian Energy, Proc. Int. Seminar, Sounion, Greece, September 1961. Plenum Press, New York, 1964.

(17) Solar Energy as a National Energy Resource, NSF/NASA Solar Energy Panel, December 1972.

(18) Report of Committee on Solar Energy Research in Australia, Australian Academy of Science, July 1973.

(19) Lalor, E., Solar Energy for Ireland, Report to the National Science Council, Dublin, February 1975.

(20) Solar Energy: a UK assessment, UK Section, ISES, London, May 1976.

(21) McVeigh, J. C., Advances in Solar Energy, *Heating and Ventilating News*, 18, 1975.

(22) Solar energy utilisation in USA, France, Italy and Australia. Proc. Conf. Brighton Polytechnic, UK Section, ISES, July 1974.

(23) deWinter, F. and deWinter, J. W., eds, Description of the Solar Energy R & D programs in many nations, ERDA Division of Solar Energy, February 1976.

(24) Eggers-Lura, A., ed, Flat plate solar collectors and their application to dwellings, Commission of the European Communities study contract No. 207-75-9 ECI DK, Copenhagen, February 1976.

(25) Sun: Mankind's Future Source of Energy. Proceedings of the International Solar Energy Congress, New Delhi, 16-21 January 1978. Pergamon Press, Oxford, 1978.

(26)   Chatel, B., Some solar energy programmes in the United Nations system, *Solar Energy* 23, 263-269, 1979.

(27)   Sun II, Proceedings of the International Solar Energy Society Silver Jubilee Congress, Atlanta, Georgia, May 1979, Pergamon Press, Oxford 1979.

(28)   Swain, H., Overend, R. and Ledwell, T. A., Canadian Renewable Energy Prospects, *Solar Energy* 23, 459-470, 1979.

(29)   Lewis, C. W., The prospects for solar energy use in industry within the United Kingdom, *Solar Energy* 24, 47-53, 1980.

(30)   UK ISES Conference (C23) on European Community Solar Housing, July 1980.

(31)   UK ISES Conference (C22) on Solar Energy Codes of Practice and Test Procedures, London, April 1980.

(32)   Stammers, J. R., The European market, UK ISES Conference (C26) on Solar energy export opportunities, London, April 1981.

(33)   New and Renewable Energy in the Arab World, United Nations Economic Commission for Western Asia, Beirut, 1981.

CHAPTER 2

SOLAR RADIATION

Radiation is emitted from the sun with an energy distribution fairly similar
to that of a 'black body', or perfect radiator, at a temperature of 6000 K.
Radiation travels with a velocity of 3×10^8 m/s taking approximately
8 minutes to reach the earth's atmosphere. The value of the solar constant
- a term used to define the rate at which solar radiation is received
outside the earth's atmosphere, at the earth's mean distance from the sun,
by a unit surface perpendicular to the solar beam - is 1373 W/m^2 with a
probable error of 1-2% (1). During the year the solar constant can vary by
±3.4%, partly due to variations in the earth-sun distance.

The earth follows an elliptical path round the sun, taking about a year for
each cycle. The earth's axis is tilted at a constant angle of 23° 27'
relative to the plane of rotation at all times. The apparent daily motion
of the sun across the sky viewed from any particular location on earth
varies cyclically throughout the year and is defined by the angle of
declination. This is the angle formed at solar noon between a vector
parallel to the sun's rays which would pass through the centre of the earth
and the projection of this vector upon the earth's equatorial plane. The
angle of declination varies from +23° 27' to -23° 27'. This affects the
angle of incidence of the solar radiation on the earth's surface and causes
seasonal variations in the length of the day. At the equator the day lasts
for exactly 12 hours from sunrise to sunset, but at higher latitudes there
is a considerable variation. For example, in the British Isles the day
lasts for less than 8 hours in mid-winter compared with 16 hours in mid-
summer. This means during the mid-summer period the total radiation on a
horizontal surface in the British Isles can be greater than in the
equatorial regions.

Four dates in the year have a particular significance. These correspond to
the two points in the earth's orbit when the effect of the earth's tilt is
at a maximum, the solstices, and the two points when the tilt has apparently
no effect, the equinoxes. In the northern hemisphere at the summer solstice,
which occurs about 22 June, the Arctic has continuous daylight as the North
Pole is at its closest position to the sun. Similarly in the southern
hemisphere at the winter solstice, about 22 December, the Antarctic
experiences continuous daylight. The sun is directly over the Tropic of
Cancer at noon at the summer solstice and directly over the Tropic of

13

Capricorn at noon at the winter solstice. At the spring and autumn
equinoxes, which occur about 21 March and 23 September, the sun is directly
over the equator at noon and, at any point on the earth's surface, day and
night last for exactly 12 hours. Tables and charts referring to
astronomical data normally relate to solar time, that is time relative to
noon with the sun at the due south position in the northern hemisphere (or
due north in the southern hemisphere). Solar time is often slightly
different from local standard time as this can apply over several degrees of
longitude and one degree of longitude is equivalent to four minutes of
standard time.

GLOBAL, DIRECT AND DIFFUSE RADIATION

The availability of solar energy in any location in the world can be studied
by two methods. The first involves measurements from a radiation monitoring
network and the second is based on the use of physical formulae and
constants. Direct solar radiation, I, is the solar radiation flux
associated with the direct solar beam from the direction of the sun's disc,
which may be assumed to be a point source, and is measured normal to the
beam (that is on a plane which is perpendicular to the direction of the sun).
Diffuse radiation, D, reaches the ground from the rest of the whole sky
hemisphere from which it has been scattered in passing through the
atmosphere. Global solar radiation, G, includes all the radiation, direct
and diffuse, incident on a horizontal plane. The distribution of diffuse
radiation is not uniform over the whole sky hemisphere and is more intense
from a zone of about 5° radius surrounding the sun. This is known as
circumsolar radiation. Radiation may also be reflected from the ground onto
any inclined surface, though this is very difficult to assess. The
relationship between direct radiation, I, the diffuse radiation, D, and the
global radiation, G, is given by

$$G = D + I \sin \gamma \qquad\qquad (2.1)$$

where γ is the solar altitude above the horizon.

SPECTRAL DISTRIBUTION OF DIRECT SOLAR RADIATION

The spectral distribution of direct solar radiation is altered as it passes
through the atmosphere by absorption and scattering. The amount of energy
absorbed depends on the length of path the solar beam traverses. A common
method of describing relative energy levels is the air mass, which is the
ratio of the actual length the solar beam traverses relative to the depth of
the atmosphere with the sun in its zenith position. Referring to Fig. 2.1.
the zenith path, ZO, is defined as unit air mass, the angle ZOS between the
zenith and the sun's direction is termed the zenith distance, z, and the air
mass, m, = SO/ZO = sec z, provided the curvature of the earth is neglected.
The second relation is very nearly equal to the true value allowing for
curvature up to 70° (air mass = 2.92). Beyond this it is necessary to allow
for variations in atmospheric refraction and to decrease of density with
height, as well as for curvature (2).

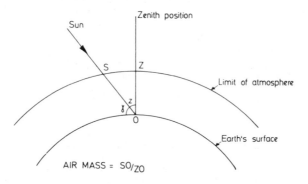

Fig. 2.1.

The spectral distribution curves for four different cases are shown in Fig. 2.2. Two are for theoretical 'black body' radiation, the first at 6000 K and the second at 5630.7 K which is the sun's equivalent black body temperature with the same overall radiation output as the solar constant (3). The third curve shows the extra-terrestrial solar spectrum (3) and the fourth and lowest curve represents the direct solar spectrum for a relatively clean atmosphere, calculated for a zenith angle of 30° and equivalent to a typical clear day in rural England at noon in summer (4). The aerosol (dust) attenuation was calculated for a continental size distribution and the Ozone and Rayleigh attenuation estimated from Elterman (5). Also from this lowest curve it can be seen that the radiation is limited to wavelengths between the near ultra-violet of 0.3 μm and the middle infrared region of about 2.5 μm. Absorption by gases and water vapour or cloud droplets occurs only in certain specific narrow wavebands. The absorption of radiation by cloud is surprisingly small, perhaps less than 10% for a cloud 1000 m thick (6), but the main loss is due to scattering. Absorption by aerosol also occurs. Scattering of radiation by cloud droplets and aerosols depends on wavelength and particle size. With low particle concentrations the scattering tends to be forward, giving relatively intense white diffuse radiation under hazy skies or thin cloud. A very dense cloud 1000 m thick could reflect back into space more than 90% of the incident solar radiation. The study of spectral distribution is based on the use of physical formulae and constants and is of immense importance in all photochemical and photobiological applications.

The peak radiation received on earth is about 1.0 kW/m^2 normal to the beam and the direct component of this is about 0.8 kW/m^2 when the sky is clear. This gives an effective depletion to about 70% of the value of the Solar Constant. It is very useful to bear this figure of 1.0 kW/m^2 in mind when assessing the performance of a solar energy system, as any claims for an output from the system which approach the value of 1.0 kW/m^2 should be treated with considerable caution.

In a general survey of solar radiation measurement, techniques and instrumentation, published in 1976, Thekaekara (7) records the adoption of 1353 W/m^2 as the solar constant during the 1970 ISES Congress in Melbourne. This value was widely used during the 1970s but was reassessed by Frolich (1) in 1977 and the revised value on the Solar Constant Reference Scale is 1373 W/m^2.

RADIATION MEASURING INSTRUMENTS

The earliest standard instruments for measuring direct beam radiation were
the Ångström pyrheliometer developed in Stockholm and the Abbot water flow
calorimeter of the Smithsonian Institute in Washington. In the Ångström
instrument the heating effect on a receiving element exposed to solar
radiation is matched by an electrically heated shaded element. Standard
methods are used to measure the electrical heating. The Abbot water flow
calorimeter contains a cavity which absorbs solar radiation and the
temperature rise in the circulating cooling water is proportional to the
solar irradiance. The Abbot silver disc pyrheliometer is a secondary
standard in which the rate of change in the disc temperature is approximately
proportional to the irradiance. For many years it was noticed that the
American and European radiation measurements did not agree, with differences
found by various investigators in many countries, ranging from between 2.5%
and 6% (8). In September 1956 a new scale, designated the International
Pyrheliometric Scale 1956, was approved which applied corrections of +1.5%
to the Ångström scale and -2.0% to the Abbot, Smithsonian scale.
Subsequently most measurements since then have been based on the
International Pyrheliometric Scale 1956. The Solar Constant Reference Scale,
which appeared in the mid-1970s, is some 2% higher than this scale.

Pyranometers, which are used for measuring global radiation, or, when shaded
from the direct beam, for diffuse radiation, are often based on the
principle of measuring the temperature difference between black (radiation
absorbing) surfaces and white (radiation reflecting) surfaces by the use of
thermopiles. These give millivolt outputs which can be conveniently handled
by a variety of conventional data recording systems and the Eppley
pyranometer is a typical example of this type. Another well-known type is
the Robitsch, based on differential expansion of a bimetallic element, while
the Bellani distillation pyranometer measures the global irradiance over a
given period of time by the distillation of alcohol with a calibrated
condenser. A much simpler type of measurement which is carried out in many
locations is the duration of 'bright sunshine'. This is measured by the
Campbell-Stokes recorder which uses a spherical lens to focus the solar
radiation onto a heat-sensitive paper which burns at the onset of 'bright
sunshine'. This can be correlated with the total global radiation by using
the regression equation

$$G = G_1 \left[a + \frac{bn}{N} \right]$$

where G is the mean global radiation on a horizontal surface; G_1, the
reference global radiation; n, mean value of the duration of bright sunshine;
N, mean length of day (or maximum possible daily value of bright sunshine);
and a and b are constants. The usual length of time considered for
applications of this formula is one month.

A good example of the way in which the equation can be applied was given by
Connaughton's (9) analysis of radiation in Ireland where the data from 23
stations recording bright sunshine was correlated with the Valentia data
from September 1954 to August 1965, giving values of $a = 0.25$ and $b = 0.58$.
A series of maps giving the estimated mean global solar radiation for each
month was then prepared. Similar work was carried out by Day (10) for the
whole of the British Isles. Day's work was more elaborate as he found that
the constants a and b varied widely from one station to another. Variations
can also occur from one period to another at the same station, as Day's
values for a and b at Valentia from 1954 to 1959 were 0.22 and 0.65,
respectively.

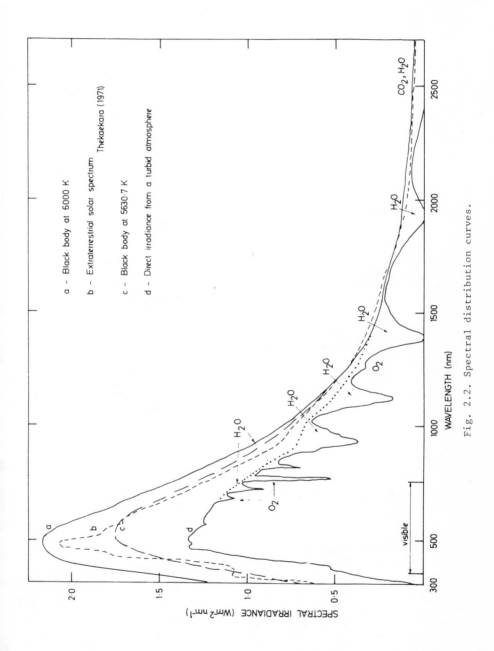

Fig. 2.2. Spectral distribution curves.

DATA FROM A RADIATION MEASUREMENT NETWORK

It is difficult to obtain reliable solar radiation data. Even experienced meteorological observers find accuracies of better than ±5% in a continuous series of long term observations hard to achieve. The most reliable data is associated with the main meteorological stations[*], but these are often widely dispersed and a considerable distance from the location of any potential application. Fortunately in the UK for most practical design purposes it can be assumed that the averaged radiation data from any meteorological station within 150 km will be perfectly adequate. There is very little systematic variation up to a distance of 300 km (11). This is illustrated by reference to Table 2.1, where it can be seen that there are two main trends - higher levels of global radiation towards the west, where the skies are, in general clearer, and lower levels towards the north, which would be expected because of the higher latitude.

This data is for the period 1965-1970, with the exception of Aldergrove which is for 1969 and 1970 only, and was derived from Meteorological Office records by Page (12).

Table 2.1. Annual variation of the mean daily totals of global solar radiation on a horizontal plane (MJ/m^2)

	Kew	Aberporth	Aldergrove	Eskdale Muir	Lerwick
January	2.13	2.39	1.67	1.54	0.82
February	4.13	5.03	4.50	4.36	2.97
March	8.06	9.51	7.29	7.30	6.41
April	11.62	14.25	12.22	11.47	12.21
May	15.54	16.89	14.65	12.99	13.60
June	18.06	20.09	19.48	16.57	16.90
July	16.03	18.13	15.35	13.64	15.42
August	13.29	15.08	13.45	12.24	11.93
September	9.73	10.58	8.86	7.79	6.88
October	5.79	6.16	4.39	4.52	3.54
November	3.00	2.97	2.66	2.41	1.37
December	1.72	1.94	1.46	1.37	0.55

More than half the solar radiation received in the British Isles is diffuse and this puts limitations on any applications which involve focusing. Figure 2.3 shows the six year average (1965-1970) of global solar radiation at Kew, London, with the direct and diffuse components. The comparatively low radiation levels in the winter period are combined with an increased proportion of diffuse radiation, which greatly reduces the effectiveness of many solar space and water-heating systems.

Applications in architectural design and housing often require a knowledge of the total radiation on an inclined surface facing in any direction, while the only available data is the total global radiation on a horizontal surface in the same location or within a reasonable distance. Very few meteorological stations give data for vertical irradiation, but this can be calculated and Fig. 2.4, based on data derived for Dublin by Cash (13), shows the effects of orientation on the ratio of vertical to horizontal irradiation.

---

[*]A complete list of the UK meteorological stations reporting hourly totals of solar radiation is given in Appendix 2.

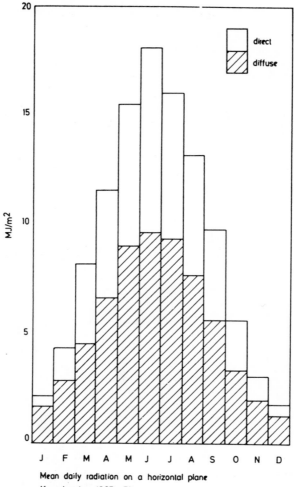

Mean daily radiation on a horizontal plane
Kew, London 1965 - 70

Fig. 2.3.

The first work on the determination of the total radiation on an inclined
surface facing in any direction in the UK was carried out by Heywood (14-16)
who suggested that calendar monthly radiation data should be replaced by a
system based on particular declination limits, having numerically equal
positive and negative magnitudes. The advantages of this system were
claimed to be as follows:

(i)   by dividing the year symmetrically about the summer solstice, periods
      of similar declination in the spring and autumn can be combined for
      the assessment of experiments or correlation with standard data;

(ii) the use of a relatively small number of standard declination values
      reduces the amount of computation;

(iii) it would provide a better basis for the comparison of radiation data.

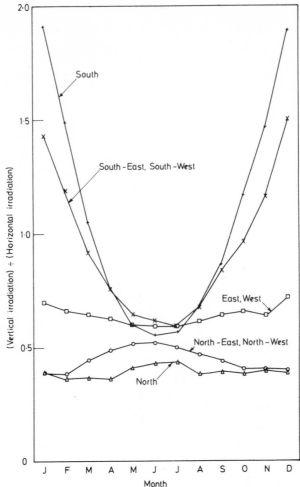

Fig. 2.4.   Effects of Orientation, Dublin derived
            data.

There was little support for this logical concept, but nevertheless Heywood
went on to establish parameters which can be determined from ratios of
radiation measurements, and produced curves showing how these could be
applied to determine the ratio of the incident radiation on any inclined
surface to the global radiation on a horizontal surface (17).   These curves
were based on experimental measurements carried out on a continuous basis
for three years at the Woolwich Polytechnic (latitude 51° 30' N).   Tables
2.2 and 2.3, based on these results, can be applied to the whole of the
United Kingdom provided that allowance is made at specific localities for
variations in the global radiation on the horizontal surface.   The 'bright
sun conditions' used by Heywood are those where the vicinity of the sun is
free from cloud and there is not more than one-third cloud cover.

Table 2.2.   Total radiation on south-facing surfaces
under bright sun conditions, MJ/m^2 day

Position of surface	Winter period Oct. 16–Feb. 26	Spring & Autumn Feb. 27–Apr. 12 Aug. 31–Oct. 15	Summer Apr. 13–Aug. 30	Annual mean
Horizontal	5.60	15.84	28.73	17.03
S 20°	9.43	20.01	30.32	20.08
S 40°	12.21	23.13	29.28	21.49
S 60°	13.48	22.78	26.16	20.67
S Vertical	12.38	18.57	16.69	15.56

Table 2.3.   Total radiation on south-facing surfaces
under average conditions, MJ/m^2 day

Position of surface	Winter period Oct. 16–Feb. 26	Spring & Autumn Feb. 27–Apr. 12 Aug. 31–Oct. 15	Summer Apr. 13–Aug. 30	Annual mean
Horizontal	2.49	7.47	14.51	8.35
S 20°	3.28	8.52	14.96	9.09
S 40°	3.79	8.99	14.50	9.20
S 60°	3.81	8.52	12.51	8.32
S Vertical	3.52	6.47	8.57	6.19

Data from Kew in the period 1959–1968 was used by the Building Research
Establishment (18) to derive the monthly and annual totals of solar energy
incident on 1 m^2 of a surface inclined at different angles to the horizontal,
shown in Table 2.4.

It can be seen that over the whole year the total radiation prediction for
angles between 30° and 60° does not vary by more than a few per cent and
that in the summer months the smaller the angle the greater the total
amount of radiation. This theoretical analysis is confirmed when Table 2.3
is examined.

An alternative approach to the problem of predicting the hourly solar
radiation incident upon any inclined surface was suggested by Bugler (19),
who used a mathematical model of solar radiation in which the diffuse
component is calculated from global horizontal radiation using three
different relationships, the appropriate equation being selected according
to the value of the ratio of measured hourly global insolation to hourly
global insolation computed for clear sky conditions. The method was checked
with reference to data for Melbourne for the period 1966–1970 with very good
results and is believed to have general application.

Table 2.4.   Monthly and annual totals of solar radiation
on inclined surfaces, MJ/m^2, (derived by BRE
computer program from Kew average solar
radiation data for 1959-1968)

Month	Direct				Diffuse*			
	30°	45°	60°	90°	30°	45°	60°	90°
January	50	65	70	70	40	40	35	30
February	70	80	85	80	65	65	55	45
March	165	180	180	145	130	130	115	95
April	170	170	160	105	190	175	165	130
May	230	215	190	105	250	240	225	180
June	250	225	190	90	265	250	235	190
July	200	185	155	75	275	265	245	190
August	210	205	185	115	225	215	195	160
September	195	205	200	150	155	145	135	115
October	135	155	160	140	100	95	85	70
November	70	85	90	90	50	45	45	35
December	50	60	70	70	35	35	30	25
Annual	1795	1830	1735	1235	1780	1700	1565	1265

*Includes ground-reflected radition.

Since the mid-1970s, many countries have greatly increased their efforts to
improve their solar radiation database.   Some examples of the type of work
undertaken range from comparisons between measured and predicted values,
basic measurements and mathematical modelling techniques as illustrated in
recent papers describing work in Greece (20), the Lebanon (21), Malaysia
(22), Nigeria (23), the UK (24, 25) and Zambia (26).

A more detailed discussion of instruments for measuring solar radiation and
techniques for handling solar radiation data is given by Duffie and
Beckmann (27), while details of International Radiation Networks and the
European Radiation Atlas are given in Appendix 3.

A very good example of this approach was shown by Sayigh (28), who used
well-known climatological data to develop an empirical formula from which
an annual total radiation curve may be calculated.   The formula includes
factors for latitude, altitude, proximity to sea or lakes, the ratio of
actual sunshine hours to the length of the day, mean relative humidity and
monthly mean maximum temperature.   This method can be applied to any
location.

An alternative approach is to select an 'example year' for the particular
country, using existing weather data.   A number of organizations in the UK,
including the Electricity Council, the National Coal Board, the British Gas
Corporation and the Building Services Research and Information Association,
have recommended that the year October 1964 - September 1965 should be
selected as an example year for comparative energy demand calculations.
This proposal has been described in detail by Holmes and Hitchin (29).

An example year can also be developed from suitable computer programs, as
shown in recent work from Italy (30).

REFERENCES

(1) Frohlich, C., Contemporary measures of the solar constant, in *The Solar Output and its Variation*, Colorado Associated University Press, Boulder, 1977.

(2) Heywood, H., Solar energy for water and space heating, *J.I. Inst. Fuel.* 27, 334-352, 1954.

(3) Thekaekara, M. P., Solar energy outside the Earth's atmosphere, *Solar Energy* 14, 109-127, 1973.

(4) McCartney, H. A., private communication.

(5) Elterman, L., Atmospheric attenuation model, 1964, in the ultraviolet, visible and infrared regions for altitudes to 50 km, Air Force Cambridge Research Laboratories, Environmental Research Papers No. 46, AFCRL-64-740.

(6) Unsworth, M. W., Variations in the short wave radiation climate of the UK, UK ISES Conference on UK Meteorological Data and Solar Energy Applications, London, February 1975.

(7) Thekaekara, M. P., Solar radiation measurement: techniques and instrumentation, *Solar Energy* 18, 309-325, 1976.

(8) Drummond, A. J. and Greer, H. W., Fundamental pyrheliometry, *Sun at Work* 3, June 1958.

(9) Connaughton, M. J., Global solar radiation, potential transevaporation and potential water deficit in Ireland, Technical Note No. 32, Department of Transport and Power-meteorological Service, Dublin, 1967.

(10) Day, G. J., Distribution of total solar radiation on a horizontal surface over the British Isles and adjacent areas, *The Meteorological Magazine* 90, 269-284, October 1961.

(11) Monteith, J. L., Contribution to discussion on (5).

(12) Page, J. K., in *Solar Energy*. Memorandum by the U.K. Section, International Solar Energy Society. Select Committee on Science and Technology (Energy Resources Sub-Committee), Appendix 1, Part 1, House of Commons Paper 156-i, HMSO, January 1975.

(13) Cash, J., Solar Energy and Buildings, Paper presented to Building Design Team, IIRS, Ireland, 5 December 1974.

(14) Heywood, H., Standard date periods with declination limits, *Solar Energy* 9, 223-225, 1965.

(15) Heywood, H., Solar radiation on inclined surfaces, *Solar Energy* 10, 46-52, 1966.

(16) Heywood, H., A general equation for calculating total radiation on inclined surfaces, Paper 3/21, International Solar Energy Conference, Melbourne, Australia, 1970.

(17) Heywood, H., Operating experiences with solar water heating, *JIHVE* 39,
      63-69, June 1971.

(18) Courtney, R. G., An appraisal of solar water heating in the UK,
      Building Research Establishment Current Paper CP 7/76, 1976.

(19) Bugler, J. W., The determination of hourly solar radiation incident
      upon an inclined plane from hourly measured global horizontal
      insolation, CSIRO, SES Report 75/4, 1975, and in *Solar Energy* 19,
      477, 1977.

(20) Pissimanis, D. K. and Notaridou, V. A., The atmospheric radiation in
      Athens during the summer, *Solar Energy* 26, 525, 1981.

(21) Sfeir, A. A., Solar radiation in Lebanon, *Solar Energy* 26, 497-502,
      1981.

(22) Chuah, D. G. S. and Lee, S. L., Solar radiation estimates in Malaysia,
      Solar Energy 26, 33-40, 1981.

(23) Ezekwe, C. I. and Ezeilo, C. C. O., Measured solar radiation in a
      Nigerian environment compared with predicted data, *Solar Energy* 26,
      181-186, 1981.

(24) Page, J. K., Souster, C. G. and Sharples, S., Mathematical modelling
      of hourly variations in temperature, wind speed and longwave
      radiation for different classes of radiation day in the UK, UK ISES
      Conference (C18) on Meteorology for Solar Energy Applications, London,
      January 1979.

(25) Munroe, M. M., Estimation of totals of irradiance on a horizontal
      surface from UK average meteorological data, *Solar Energy* 24, 235-238, 1980

(26) Lewis, G., Irradiance stimates for Zambia, *Solar Energy* 26, 81-85, 1981.

(27) Duffie, J. A. and Beckmann, W. A., *Solar Engineering of Thermal
      Processes*, pp. 1-110, John Wiley, New York, 1981.

(28) Sayigh, A. A. M., Estimation of total radiation intensity, *Jl Engng
      Sc*, University of Riyadh, Saudi Arabia, Vol. 5, No. 1, pp. 43-55, 1979.

(29) Holmes, M. J. and Hitchin, E. R. An "example year" for the
      calculation of energy demand in buildings. Building Services
      Engineer, 45, 186-189, 1978.

(30) Ahdretta, A., Bartoli, B., Coluzzi, B., Cuomo, V., Francesca, M. and
      Serio, C., A computer program to determine the reference year,
      Paper H3:22, ISES Solar World Forum Abstracts, Pergamon Press, Oxford
      1981.

CHAPTER 3

WATER AND AIR HEATING APPLICATIONS

Solar energy can be easily converted into heat and could provide a significant proportion of the domestic hot water and space heating demand in many countries. One of the drawbacks in high latitude countries, such as the UK, is that there are many days in the winter months when the total radiation received will be too small to make any useful contribution. The most widely known and understood method for converting solar energy into heat is by the use of a flat plate collector for heating water, air or some other fluid. The term 'flat plate' is slightly misleading and is used to describe a variety of different collectors which have combinations of flat, grooved and corrugated shapes as the absorbing surface, as well as various methods for transferring the absorbed solar radiation from the surface of the collector to the heated fluid. Many different types of collector have been built and tested by independent investigators over the past ninety years, the early work being carried out mainly in the USA (1, 2), the UK (3), Australia (4), South Africa (5) and Israel (6). Tests were carried out in specific locations with wide variations in test procedures and in the availability of solar radiation. The main objective of these tests has been to convert as much solar radiation as possible into heat, at the highest attainable temperature, for the lowest possible investment in materials and labour (7).

The major British research in this field was carried out by the late Professor Harold Heywood, commencing with experimental work in 1947 on the characteristics of flat plate collectors (3). His earliest experiments were carried out on a small square collector, 0.093 m^2 in area. The heat collected was absorbed by means of water channels soldered onto the back surface of a blackened copper plate and the rate of heat absorption was determined for various numbers of glass plates and for different temperatures of collection. The somewhat simplified theoretical treatment which he established at that time was used as the basis for some of the design work on domestic flat plate collectors carried out in the 1970s.

As well as these fundamental studies on the principles of heat collection
Heywood built a solar collector of approximately 1 m^2 area which worked
satisfactorily for many years in his home about 15 km south-east of London.
The collector was constructed from two sheets of corrugated galvanized iron
and was installed in a conventional thermosyphon system. The water capacity
of the collector was 22.5 l and the storage tank had a similar capacity,
giving a total water capacity of about 45 l for 1 m^2 of collector surface.
This particular ratio of collector size to water capacity has featured
prominently in subsequent experiments carried out in many different
countries. Heywood's general conclusions, which are still very relevant,
were as follows:

   (i)   Simplicity of construction must be an essential feature of water and
         space heating.
  (ii)   There is a considerable variation in the heat recovery rate from
         day-to-day in the UK.
 (iii)   Satisfactory heat collection efficiencies are only obtained where
         there is prolonged and intense direct radiation. Cloudy periods cause
         a serious reduction in the collection efficiency and, while diffuse
         radiation can be partly effective, it has much less value than direct
         radiation.
  (iv)   There is no direct correlation between 'sunshine hours' as registered
         in many parts of the UK and heat recovery.

He also commended that the variation in solar radiation, which is rarely
identical for even two or three consecutive days, made experimental work
exasperating!

In southern Florida, USA, during the late 1930s, solar energy was the main
method used for providing hot water services to single-family residences,
blocks of flats and other small commercial buildings. A survey carried out
by Scott (8) showed that nearly all the systems relied on the natural
thermosyphon principle (see Chapter 10) and the collectors consisted of
copper tubes soldered onto copper sheets, painted matt black and enclosed in
a galvanized steel casing. Evidence obtained from suppliers and users
confirmed that the collectors, considered alone, were very durable and some
had lasted completely trouble-free for over 30 years. Even freezing
conditions, encountered rarely in the Miami area, failed to damage collectors
made with soft copper. Users who had discontinued using their solar systems
did so for three main reasons:

   (i)   Damage caused by leaking main storage tanks.
  (ii)   Insufficient hot water.
 (iii)   The considerable expense required to replace the storage tanks.

The problems relating to the tanks were caused by the combination of copper
tubing in the collectors and the steel storage tanks, leading to corrosion.
The increasing use of domestic hot water through the progressive introduction
of washing machines and dishwashers meant that many systems could no longer
cope with the demand. This early experience has proved very valuable in
subsequent collector and system design studies.

Considerable practical experience was gained in Australia during the 1950s (4)
and the Commonwealth Scientific and Industrial Research Organization (CSIRO)
subsequently published a guide to the principles of the design and
construction and installation of solar water heaters (9). At that time they
observed that the solar water heating industry had become established in
Australia and it was a practical and acceptable way of providing a domestic

water supply at a reasonable cost. Straightforward items of equipment developed by CSIRO and others could be bought as standard items from suppliers throughout Australia. The first cost was greater than that of conventional installations but operating and maintenance costs were much lower. The Australian research also showed that a simple solar installation could provide adequate hot water for the needs of an average family throughout the year, although it was more convenient, and in several places more economical, to boost the solar heat with conventional heat sources. Many commercial companies started to manufacture and supply solar water heaters at that time but very few survived - mainly because there was little demand for 'off-the-shelf' solar equipment. The few that managed to continue commercial operations did so by offering complete systems and by the mid-1970s were established as leaders in the field with new designs based on years of practical experience.

In the UK the solar industry developed very rapidly from only two companies active in 1973 to over 60 by the end of the decade, with an estimated turnover of well over £5 million. Factors influencing this development were examined by myself (10) in 1979 and reassessed by Cross and Johansson (11) in 1981. Most industrial 'western' countries had at least a handful of solar companies by this point and several, such as France, Germany, Italy, Japan and the USA, had a substantial number of large multinational companies involved in solar activities. In the Arab countries, solar industries were reported in Jordan, Lebanon, Tunisia and the United Arab Emirates (12).

THE FLAT PLATE COLLECTOR

The majority of flat plate collectors have five main components, as shown in Fig. 3.1. These are as follows:

 (i)   A transparent cover which may be one or more sheets of glass or a
        radiation-transmitting plastic film or sheet.
 (ii) Tubes, fins, passages or channels integral with the collector
        absorber plate or connected to it, which carry the water, air or
        other fluid.
 (iii) The absorber plate, normally metallic and with a black surface,
        although a wide variety of other materials can be used, particularly
        with air heaters.
 (iv) Insulation, which should be provided at the back and sides to
        minimize the heat losses.
 (v)   The casing or container which encloses the other components and
        protects them from the weather.

Components (i) and (iv) may be omitted for low temperature rise applications, such as the heating of swimming pools. Some of the very great variety of solar water and air heaters are illustrated in Fig. 3.2. Corrugated, galvanized sheet is a material widely available throughout the world and Figs. 3.2(a) and (b) show two ways in which it has been used. The use of conventional standard panel radiators (5, 13), shown in Fig. 3.2(c) is one of the simplest practical applications (see Chapter 10). Methods of bonding and clamping tubes to flat or corrugated sheet are shown in Figs. 3.2(d) and (e) while Fig. 3.2(f) is the 'tube-in-strip' or roll bond design, in which the tubes are formed in the sheet, ensuring a good thermal bond between the sheet and the tube. An effective commercially available collector is shown in Fig. 3.2(g), based on standard refrigeration heat exchanger practice. Two different types of solar air heater surface are shown in Figs. 3.2(h) and (i).

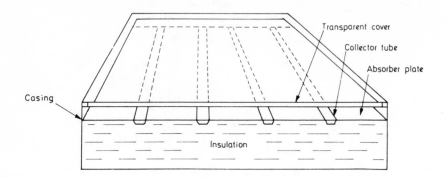

Fig. 3.1.

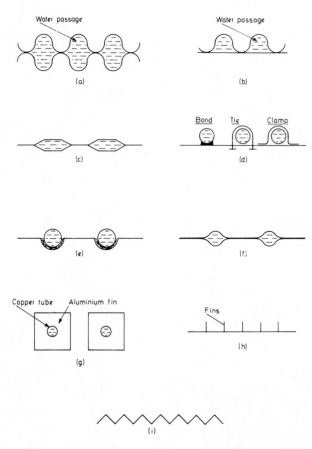

Fig. 3.2.   Cross-sections through collector plates.

Flat plate collectors could also be classified into three groups according to their main applications as follows:

(i)   Applications with a very small rise in temperature, such as in swimming pools where the collector needs no cover or insulation at the back or sides (14). A high rate of flow is maintained to limit the temperature rise to less than 2°C.
(ii) Domestic heating and other applications where the maximum temperature required is no more than 60°C. Insulation at the back and at least one transparent cover are necessary.
(iii) Applications such as process heating or the provision of small scale power, which temperatures considerably above 60°C are necessary. A more sophisticated design approach is needed to reduce heat losses from the collector to the surroundings.

From the great variety of successful collectors shown in Fig. 3.2, the flat plate collector appears to be a comparatively straightforward piece of equipment and in an ideal collector all the radiation reaching it would be converted into heat. In practice, the useful heat collected, Q, is always less than the incident solar radiation, G_c. There are many different factors which can contribute to this and a detailed analysis of the thermal characteristics of the flat plate collector is very complex. For example, the loss of heat by radiation increases as the fourth power of the absolute temperature, making losses due to radiation increasingly significant as the temperature of the heated fluid becomes more than 25°C above the temperature of the surroundings. The first detailed analysis of these various factors was carried out by Hottel and Woertz in 1942 (2). However, a comparatively simple analysis will give very useful results and show how the important variables are related and their effect on collector performance.

THE HOTTEL–WHILLIER–BLISS EQUATION

The basic equation, widely known as the Hottel–Whillier–Bliss equation (15–20), expresses the useful heat collected, Q, per unit area, in terms of two operating variables, the incident solar radiation normal to the collector plate, G_c, and the temperature difference between the mean temperature of the heat removal fluid in the collector, T_m, and the surrounding air temperature, T_a, as follows:

$$Q = F\left\{(\tau\alpha)G_c - U(T_m - T_a)\right\} \qquad (3.1)$$

where F is a factor related to the effectiveness of heat transfer from the collector plate to the heat removal fluid. This factor is influenced by the design of the collector plate, for example the dimensions of the passages containing the heated fluid and the thickness of the plate, as well as the properties of the fluid. It also varies with rate of fluid flow through the collector.

The transmittance–absorptance product $(\tau\alpha)$ takes account of the complex interaction of optical properties in the solar radiation wavelengths (20). It is actually some 5% greater than the direct product of the transmittance through the covers, τ, and the collector plate absorptance, α, because some of the radiation originally reflected from the collector plate is reflected back again from the cover.

The heat loss coefficient, U, rises very rapidly with increasing wind velocity if there are no covers, but is less dependent when the collector has at least one cover. The number and spacing of the covers and the conditions within the spaces can be significant, for example an evacuated space greatly reduces heat losses. The longwave radiative properties of the collector plate and covers also influence the heat loss coefficient.

These three design factors, F, $(\tau\alpha)$ and U, define the thermal performance of the collector and the overall efficiency of the collector, $\eta = Q/G_c$, can be expressed in terms of the temperature difference, $(T_m - T_a)$, and the incident solar radiation, G_c, in equation (3.2):

$$\eta = \frac{Q}{G_c} = F\left\{(\tau\alpha) - \frac{U}{G_c}(T_m - T_a)\right\} . \qquad (3.2)$$

The temperature T_m is almost impossible to measure, but as most systems have comparatively small temperature rises through any individual collector, the inlet fluid temperature T_i can usually be substituted for T_m. Typical values for F are about 0.88 to 0.90, $(\tau\alpha)$ for two covers of 3 mm window glass and an α value of 0.90 is about 0.7, while U for the same collector would be about 3.6. An unglazed, uninsulated collector would have a $(\tau\alpha)$ value approaching unity, but a U value at least twice that of the glazed collector. Experimental methods for determining F, $(\tau\alpha)$ and U are given by Smith and Weiss (20).

PHYSICAL DESIGN CHARACTERISTICS

The three design factors are all affected by the actual physical design characteristics of the collector, the main characteristics being the type and number of transparent covers and the properties of the collector surface. The wavelength range of the incoming solar spectrum is less than 3 μm for all than about 2% of the total incoming solar energy outside the earth's atmosphere. When this radiation reaches a sheet of glass, as much as 90% can be transmitted directly, the remainder being reflected or absorbed by the glass. The absorbed energy raises the temperature of the glass so that, in turn, the glass re-radiates from both the internal and external surfaces. As the surface temperature of the collector plate rises, it also radiates, but at greater wavelengths than 3 μm for all but a very small proportion of the total energy, typically less than 1% for a black surface at $100°C$. The longwave radiation emitted by the collector plate cannot pass back directly through the glass, as the transmittance of glass is practically zero in the range 3 - 50 μm. This phenomenon is the well-known 'greenhouse effect' and the use of a transparent cover or several covers greatly reduces heat losses from the collector. Transparent plastic materials also possess high shortwave transmittance characteristics, but generally have appreciable longwave transmittances. With direct radiation the transmittance varies with the angle of incidence as shown in Fig. 3.3, where transmittance values for single and double glazing using double-strength clear window glass (21, 22) are compared with a fibreglass plastic material (23). This particular material has exceptionally good properties in the longwave region, as shown in Fig. 3.4.

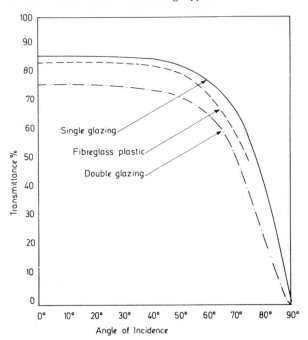

Fig. 3.3.

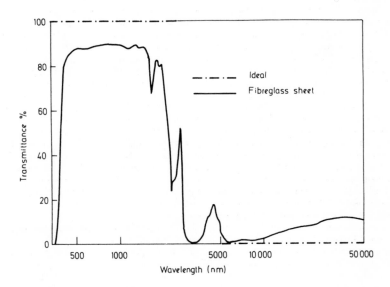

Fig. 3.4.   Spectral transmittance, Kalwall fibreglass sheet.

Each transparent cover reduces the outward heat losses from the front of the collector, but also reduces the total amount of incoming solar radiation which can reach the collector plate surface. When the absorption of energy into each cover is taken into account, the transmittance losses for incidence angles up to 35° are approximately 10% for single glazing, 18% for double glazing and 25% for triple glazing (22). The use of a combination of an outer glass cover with an inner, cheaper transparent plastic film can have advantages, as the plastic can have a higher transmittance than glass and the outer glass cover provides a certain amount of protection from weathering (24). The distance between covers or between the inner cover and the collector plate surface is not very critical (19). An optimum spacing of between 10 to 13 mm has been suggested (25), but up to 25 mm can be used.

The performance of a glass cover can be improved by depositing a transparent coating on its inner surface which allows nearly all of the incident solar radiation to be transmitted, but reflects any longwave radiation back to the emitting surface of the collector plate. Indium oxide and tin oxide are commonly used and a Japanese vacuum formed coating (26) gave a transmittance of 0.85 in the visible range (0.55 μm) and a reflectance of about 0.97 in the infra-red (4.0 μm). The early figures quoted for the Philips evacuated tubular collector (27) gave values of 0.85 and 0.9, respectively for τ and α. α.

A somewhat unexpected hazard to solar collectors, possible damage from an intense hailstorm, was reported by Löf and French (28). During an intense hailstorm in Fort Collins on 30 July 1979, hailstones of golfball size and larger caused property damage estimated at $50 million. Of ten solar heated buildings in the main hail path equipped with tempered glass covers, six suffered no damage. The general conclusion was that the risk of hail damage to commercial solar collectors glazed with 3 mm tempered glass of 0.6-1.1 m width, mounted at slopes greater than 30°, is negligibly small. A statistical model for estimating the risk of impacts by large hailstones on any ground installation, such as a solar collector array, has also been developed (29).

SELECTIVE SURFACES

The longwave radiation emitted from the collector plate surface can be considerably reduced by treating the collector surface. The treatment reduces its emissivity in the longwave spectrum without greatly reducing the absorptivity for shortwave radiation. This concept is shown in Fig. 3.5, where the properties of an ideal selective surface are illustrated. The monochromatic reflectance is very low below wavelengths of 3 μm, known as the cut-off or critical wavelength, and very high above this value. For the great majority of flat plate collectors the temperature of the surface will be sufficiently low for practically all the emitted energy to occur in wavelengths greater than 3 μm. The contrast between the ideal surface and some real surfaces is shown when Fig. 3.6, adapted from McDonald (30), is examined. Real selective surfaces do not show the sharp rise in reflectance at one particular cut-off wavelength and their properties vary with wavelength. Complete integration over the emitted spectrum is necessary to estimate the long wavelength emittance and over the solar spectrum to estimate the solar absorptance.

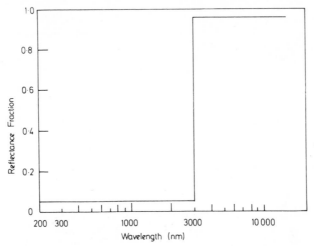

Fig. 3.5.   Spectral reflectance of ideal surface.

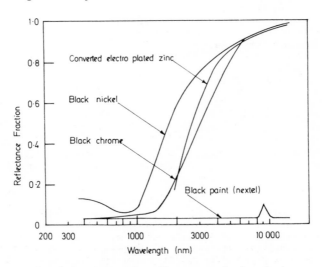

Fig. 3.6.   Spectral reflectance of plated zinc.

The effect on the performance of the collector of increasing the number of
covers and applying a selective surface is shown in Table 3.1, which gives
the overall loss to the surroundings from the top cover, assuming a mean
wind velocity of 5 m/s and an ambient temperature of 10°C. The collector
plate is considered at 40°C, a typical summer temperature achieved in the
British Isles, and at 80°C, a temperature suitable for many process heat
applications. Back and side losses from the collector casing are not
considered. The figures are based on data from Duffie and Beckman (19).

Sun Power

Table 3.1. Overall top cover loss (W/m^2)

Plate temperature	$40^{\circ}C$		$80^{\circ}C$	
Longwave emissivity	0.95	0.1	0.95	0.1
1 cover	189	93	525	263
2 covers	78	57	280	168
3 covers	63	45	182	119

The reduction in energy loss due to the selective surface becomes increasingly significant as the temperature of the collector plate rises. Any reduction in energy loss improves the collector efficiency and the overall annual increase in collected energy which can be achieved by the use of selective coatings depends on the number of hours per year during which there is sufficiently intense radiation to allow the collector to reach temperatures at which this becomes significant. An increase of up to 20% in the amount of energy collected in a year in the British Isles has been suggested (31).

The use of a second cover has almost the same effect on the top cover loss as a good selective surface in this temperature range, but this also reduces the amount of radiation reaching the collector plate surface. At comparatively small temperature differences between the collector plate surface and the surroundings, a single glazed collector is normally more efficient for this reason. The use of a selective surface and two covers give a comparatively small improvement over the selective surface with one cover.

SURFACE TREATMENT

There are several different ways in which selective surfaces can be prepared, depending on the physical principle involved. Tabor (32) has been concerned with finely divided metals deposited on polished metal undersurfaces; other methods include the deposition of thin semiconductor layers which absorb the short wavelengths but not the long, thus allowing the metallic undersurface to maintain its low emissivity. Surfaces may also be given a controlled degree of roughening so that only the absorptivity for short wavelengths is increased. Surfaces with large (relative to all radiation wavelengths) V-grooves can be arranged so that radiation from near normal directions to the overall surface will be reflected several times in the grooves. An effective α/e of 9 with $\alpha = 0.9$ has been suggested (19).

A method for the commercial production of a selective surface on a copper surface was described by Close (33). It consisted of dipping the copper plate in a solution of 1 part sodium chloride $(NaClO_2)$ to 2 parts sodium hydroxide (NaOH) in 20 parts of water by weight. The plate should be immersed for 10 minutes at a controlled temperature of about $62^{\circ}C$. The usual precautions of ensuring that the plate was clean and degreased prior to immersion were recommended. The performance of a single glazed collector treated with this surface was found to be about 10% better in overall collection efficiency compared with a conventional non-selective double glazed collector.

Some chemically applied coatings referred to in the literature include a
nickel-zinc-sulphide complex known as 'nickel black' (32), copper oxide on
copper (32, 34) and copper oxide on aluminium (35) and a black chrome
coating, the Harshaw ChromOnyx® process (30, 36, 37). This is a modified
version of a well-known standard decorative black chrome electroplate seen,
for example, on office furniture. A comparison between black chrome
processes and other solar selective coatings is given in Table 3.2.

Table 3.2

Coating (and reference)	Absorptance (α)	Emissivity (e)	Ratio of absorptance to emissivity (α/e)
Nickel black on galvanized iron (experimental) (32)	0.89	0.12	7.42
Same process (32)	0.89	0.16-0.18	5.56-4.94
Sodium hydroxide, sodium chlorate on copper (33)	0.87	0.13	6.69
Black chrome on dull nickel (36)*	0.923	0.085	10.86
Black chrome on bright nickel (36)*	0.868	0.088	9.86
Black nickel (36)*	0.867-0.877	0.066-0.109	7.95-13.29
Nextel black paint (36)*	0.967	0.967	1.00

*Results based on spectrum weightings for solar air mass 2 (absorptance) and
121°C black body (emissivity).

A selective surface with a high ratio of absorptance to emissivity
(α/e = 20), with α nearly 1, has been obtained with a gold black coating
(38, 39) placed on a reflecting undersurface such as copper. A relatively
inexpensive electrochemical method which uses a chromium based oxide coating
known commercially as Solarox has been developed in Australia (40). Typical
results gave an α/e value of 18 at 25°C falling to 7.5 at 300°C.

An alternative method of obtaining a selective surface on an absorber plate
is to bond a thin, selectively coated nickel foil onto the absorber. The
surface durability, which is one of the major problems with selective
surfaces, is very good, with no deterioration observed after three years
exposure. Other advantages claimed for selective foils compared with
conventional batch processing are:

(i)   high uniformity and consistency of the surface;
(ii) no limitation on collector size and choice of substrate material;
(iii) minimizes collector plate handling and transportation costs and
      provides flexibility in installation procedures.

COLLECTOR MATERIALS AND CORROSION

All components and materials used in a solar energy collector should be
designed to operate satisfactorily under the worst conditions which could be
expected in any particular installation. Materials should be capable of
withstanding both the high temperatures which would be encountered during
periods of maximum radiation with no flow through the collector and the low
temperatures which could occur in mid-winter. Problems which could arise
from cyclic variations in temperature or large temperature differences
within the collector should be taken into consideration in materials
selection and design. The estimated life of any component is important in
determining the real cost of the delivered energy and corrosion may be the
greatest limiting factor.

The majority of solar heating systems have more than one metal in contact
with the heat transfer fluid, which is usually water. Pipework may be
copper or stainless steel and the collector plate could be constructed from
copper, stainless steel, mild steel or aluminium. The presence of mixed
metals in a system is one of the most important mechanisms which can give
rise to corrosion. The other is the presence of dissolved oxygen in the
heat transfer fluid (42). In the simplest solar water heating system, the
water from the cold tank passes through a solar collector to preheat the
feed to the hot system. Oxygen is freely available in such a system and the
wrong choice of materials for the pipework and collector could result in a
very short life before perforation occurs. For example, an aluminium
collector panel tested with a steady flow of ordinary mains water containing
dissolved copper failed in less than two months (43). There are also the
further problems of scale formation in hard water areas and possible frost
damage in winter.

In closed circuit systems a heat exchanger is placed in the solar hot water
storage tank and the water recirculates through the collector and heat
exchanger. Standard finned copper tube can be used as the heat exchanger
and an experimental unit tried at Brighton Polytechnic in the early 1970s is
shown in Fig. 3.7. This concept has now appeared in several commercial
systems.

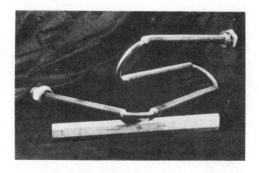

Fig. 3.7.

These systems usually rely on initial corrosion to reduce the oxygen content in the water to an acceptable level. A pump is often incorporated in the circuit and the wrong positioning of the pump could send water up through the expansion tank, picking up fresh oxygen. Ideally the expansion tank should be sealed. Bacterial activity can also cause problems as some antifreeze solutions when heated provide ideal conditions for microorganism growth, especially if the mains water used to charge the system contains dissolved salts. Biocides can reduce this activity and the dissolved salts could be eliminated by using deionized water. Mixed-metal systems allow maximum use to be made of less-expensive materials such as uncoated mild steel, but they can only be used safely in a closed circuit with the addition of a suitable inhibitor to prevent cuprosolvency and a biocide. If glycol is added for frost protection it should be a suitably inhibited grade. The use of a sealed closed circuit may be more acceptable to some local water authorities who would object to the use of inhibitors with systems directly connected to a mains water supply.

The estimated useful life of a selective surface is hard to assess and the initial values of both α and e may degrade with use. Measurements which have been made on some selective surfaces have shown that e increases with long-term use (44). Two possible causes are suggested for this, ultra-violet radiation or the effects of atmospheric moisture and pollution. The insulation in any collector should have a low thermal conductivity and be thermally stable at the maximum collector temperature. The various materials assembled in a collector affect its thermal performance, but these effects are not independent and must be evaluated in each specific application. The temperature level at which the collectors would be operating is particularly significant, for example, a selective surface could be shown to be cost effective in one case but not in another.

A comprehensive review of the problems associated with the use of aluminium and copper was given by Popplewell (45). It includes a discussion of system design to avoid corrosion and presents corrosion data in fresh water for various copper alloys. The use of an organic, non-corrosive fluid as the heat transfer medium was considered to be an acceptable alternative.

Problems of aluminium/copper corrosion have been eliminated in the advanced absorber plate developed by Aluminiumteknik in Sweden (46). Their basic element consists of an aluminium strip with an integrated copper tube which is metallurgically bonded to and completely enclosed by the aluminium strip. The heated fluid is only in contact with the copper while the thickness of the aluminium fins and the tube can be chosen independently. As a result the strip weight is only 2.2 kg/m^2. The plate is produced by a continuous rolling process, which naturally leads to lower costs with increasing production. A selective surface has also been developed ($\alpha = 0.95$ and $\varepsilon = 0.15$) which is claimed to cost about half as much as black chroming. Accelerated and field corrosion tests indicate a service life of well over 15 years. This is another 'second generation' absorber with the added advantage that the unit costs will be considerably lower than most of the first generation, 1970s, absorbers.

DEVELOPMENTS IN COLLECTOR DESIGN

The overall thermal efficiency of a collector, which was derived from the Hottel-Whillier-Bliss equation in equation (3.2), can be used to compare the performance characteristics of different types of collector and to explore the effects of altering the various design parameters.

It must be emphasized that the equation is approximate and that at higher
values of ΔT the radiation losses from the collector become significant.
As these are proportional to the fourth power of temperature, the real
characteristic becomes curved. Radiation losses also increase if the
incident radiation increases, as a clear sky has a slightly lower effective
temperature than a cloudy atmosphere or a sky with a higher diffuse
radiation content. An increase in the wind speed across the collector
surface reduces the overall performance. The effect of double glazing,
compared with single glazing, is to improve the efficiency at higher values
of $(T_m - T_a) \times G_c^{-1}$, as heat losses are reduced, but to lower it at low
values, because the incoming radiation reaching the absorber plate is less
as it passes through two covers. This is illustrated in Fig. 3.8, which is
based on data originally presented by Justin (47).

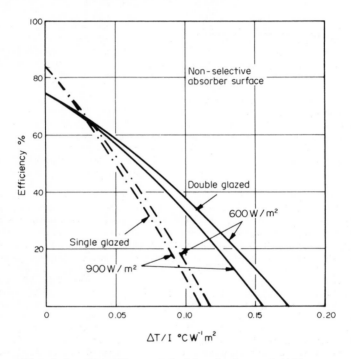

Fig. 3.8. The effect of double glazing and different
           levels of insolation.

There are many different approaches to solar collector design and it is only
possible to examine some of the more interesting trends in detail.
Improvements in the overall efficiency of collection, particularly with
higher temperature differences between the heated fluid and the surroundings,
can seldom be achieved without increasing the complexity and cost of the
collector. For low temperature rise applications the emphasis is on designs
which can be shown to have very short payback periods (the capital cost of
the system divided by the current annual value of the fuel saved) in the
order of five years or less.

Low Temperature Rise Applications

At present the main application is in swimming pool heating, but there are many other potential applications for this type of collector, such as in the glasshouse industry and in fish-farming. The cheapest, simplest and most direct method of heating any outdoor swimming pool is by the direct absorption of the incident solar radiation on the surface of the pool. With no form of swimming pool cover or any other means of preventing heat loss from the water surface the temperature of the pool in a temperate climate such as in the British Isles would normally follow the mean air temperature fairly closely during the summer months. However, the summer swimming season can be extended by one or two months at either end by providing a solar water heating system in addition to the direct absorption mentioned above. Another extremely important factor is the reduction of the heat losses from the pool. The heat loss by evaporation is the most significant (48), but fortunately this can be almost completely eliminated by the use of a single thin cover on the surface. Experiments in Australia (49) and at the University of Florida (50, 51) indicated that the use of a floating transparent plastic cover could raise the average pool temperature by over 5°C compared with a similar identical unheated pool. A smaller temperature rise can be obtained in the British Isles, partly due to the poorer radiation climate and partly to the greater rainfall causing the cover to be partially submerged and reducing its effectiveness. The other main losses are through convection and radiation. Pools placed with exposed walls above ground level are usually colder than conventional pools. Conduction losses from conventional pools can be neglected as practically all the heat that escapes into the ground returns again to the pool when the pool temperature falls (50).

For these low temperature-rise applications, where the temperature rise is only 1 or 2°C to reduce heat losses from the collector, a simple unglazed uninsulated collector is quite adequate and several successful commercial designs using different plastics materials are available. Some designs have been based on the use of black corrugated sheets with water flowing down the channels from a perforated pipe. Known as the 'trickle' type, they have been used extensively in the USA by Thomason, and some applications are described in Chapter 4. Various types of black sheeting could be placed over the corrugations or a single galvanized sheet can be painted black and wrapped in transparent plastic, with the water trickling down both the front and the back of the sheet (52). For high efficiency a uniform thin film of water is desirable and a method for achieving this is described in Chapter 10 where the black sheet is placed over a sheet of polythene packing material containing a uniform matrix of equally spaced cylindrical air bubbles. The water flows between the two sheets. If higher outlet temperatures are required a transparent cover is needed; Fig. 3.9 shows an experimental collector covered with such a cover. Several commercial manufacturers have now adopted these covers as their standard material for both low and medium temperature rise applications.

In the UK, solar heated swimming pools have become accepted on economic grounds as a result of simple tests on a number of installations (53).

Fig. 3.9.

Flat Plate Collectors

The thermal trap collector. This system was first proposed by Cobble (54) and has been developed by the New Mexico State University (55). It uses a transparent solid (methyl methacrylate) adjacent to the conventional flat collector plate, as shown in Fig. 3.10. Methyl methacrylate has a high transmittance in the visible and shorter infra-red spectrum combined with a very low transmittance at longer wavelengths and a small thermal conductivity. Comparative tests carried out at New Mexico State University showed that the thermal trap collector had superior characteristics to a conventional flat plate collector and a trickle type collector. All three collectors were tested at operating temperatures between 38 and 80°C and in this range the thermal trap collector had a higher collection efficiency and could operate usefully for more hours in the day. It was affected less by intermittent cloud conditions as it had a relatively large time constant and it appears to be very promising for use as a high temperature collector.

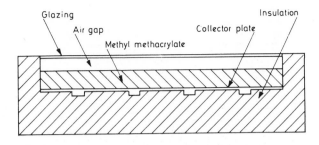

Fig. 3.10.   Thermal trap collector.

A subsequent analysis by Samuel and Wijeysundera (56) in 1980 examined both single and multilayer thermal trap collectors and showed that the number of slabs of methyl methacrylate, in addition to thickness, are important parameters.

Honeycomb systems. The use of honeycomb cellular structures placed between the transparent cover and the collector plate has been shown to be an effective method for improving the overall collector performance by suppressing natural convection losses and greatly reducing the infra-red reradiation losses. The cellular material should have a low thermal conductivity to reduce conduction heat losses from the collector plate to the outer cover. Theoretical studies (57) have predicted that a thin-film transparent plastic honeycomb could increase collector efficiency to greater than 60% at a mean collector temperature of 365 K compared with a measured 43% with a conventional double-glazed collector with a selective surface. It was considered that this could be achieved without an increase in collector cost as only one transparent cover is required for a honeycomb system. Tests carried out on a polyethylene square-cell array (58), which had 25.4 mm square sides and a depth of 76.2 mm, showed that natural convection losses had been effectively suppressed with the collector and the honeycomb in an inclined position.

The University of California, Los Angeles, are strong supporters of glass as the cellular structure material (59, 60) as it is inexpensive and readily available with low thermal conductivity. Its optical properties are excellent, as it has a very low solar absorptance and the transmitted and reflected components of direct solar radiation are specular, allowing the radiation to maintain its direction towards the collector plate. For a cellular structure consisting of an array of circular tubes the main design parameters are the internal diameter, which must be less than 150 mm, and the length, which should be less than four times the diameter. Other cellular materials under consideration have reflecting surfaces, but if these are metallized, the coating must be very thin to reduce the conductive heat loss. Earlier work had been confined to horizontal testing but later work first extended the tilt up to 30° from the horizontal (61) and then to 90° from the horizontal with various ratios of length, width and depth using honeycomb cores made from polyurethane-varnished paperboard (62).

Structurally integrated collectors. In any new installation, or in an existing building where the roof has to be replaced, considerable economic advantages can be obtained if the solar collector is combined into a structural unit so that the collector is also the roof of the building. The design criteria established by the Los Alamos Scientific Laboratory (63) included good thermal performance, economical in large-scale production, the

use of economical and readily available materials, a long service life and
capable of being easily installed and maintained by local builders. The
main features of the collector are shown in Fig. 3.11. The collector
surface was formed by welding thin mild steel plates with seam welds around
the periphery and central spot welds. The plates were then expanded under
pressure to form the flow channels. The lower extended surface of the
collector plate has three bends so that it forms a structural channel. The
upper extended surface is positioned at right angles to the collector plate,
so that adjacent units can be easily connected at the upper edge with a
U-shaped cap strip. The two glass covers are set into the trough section
and supported at the edges by a neoprene or silastic support. Each unit is
about 610 mm wide and from 2.4 to 6.1 m long. As well as the structural
advantages already mentioned, the cap strip with its compression seal
reduces the time which would otherwise be spent on sealing manually on the
site. The foam insulation also increases the rigidity of the panel. Glass
is the cover material and among the reasons given for the use of glass
rather than plastic are the problems of sealing and expansion compensation
which could arise with the relatively higher coefficients of thermal
expansion in plastics materials. A full series of test results, including
the effects of atmospheric corrosion, internal corrosion and materials
stability were presented at Los Angeles in 1975 (64).

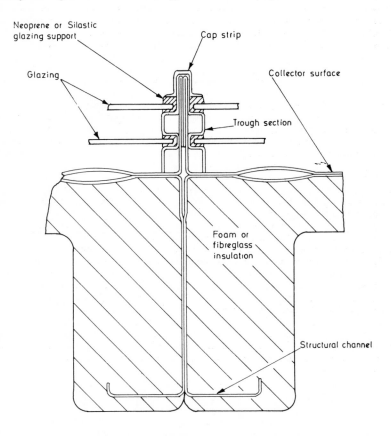

Fig. 3.11.   Structurally integrated collector.

Negative pressure distributed flow collectors. A collector design which has
overcome the necessity for the water channels to withstand a positive
internal pressure has been developed at the University of Iowa (65). The
design concept is based on flow between two parallel plates and uses either
corrugations or other forms of surface indentations on one or both sheets,
or a porous spacer, such as a screen wire, between the two sheets. Flow
through the collector is downward and at sub-atmospheric pressure, the whole
of the under surface of the collector plate being exposed to the heated
water. A significant improvement in collector performance was shown when
compared with some commercially available conventional collectors. At an
estimated temperature difference of $52.5^{\circ}C$ above the surroundings with an
incident radiation of 750 W/m^2, the negative pressure distributed flow
collector had an overall efficiency of 44% compared with 38.4% from a
conventional collector. The material for the distributed flow collector
plate could probably be copper sheet of 1.27 mm thickness, which would give
adequate strength to withstand the compressive force caused by the
atmospheric pressure-fluid pressure differential. The use of such thin sheet
copper considerably reduces the estimated materials cost of these collectors
in production when compared with conventional collectors.

A full account of the experimental work following the original concept,
which dates from the mid-1970s, was published in 1980 (66). This showed
that the mean thermal efficiencies were fairly similar to those of a simple
single-glazed flat plate collector and that the significant improvement in
performance shown in the early tests was not present in these later studies.

Some Non-Tracking Reflecting and Concentrating Systems

For most practical solar power schemes the solar radiation has to be
concentrated by a factor of about ten or more in order to achieve high
temperatures. This can be done by various tracking systems, but it would
be a considerable advantage if the required concentration could be achieved
by a stationary collector.

The compound parabolic concentrator (CPC). An important class of
concentrator, originally called the ideal cylindrical light collector, was
announced in 1974 by Winston (67). The development had its origins in the
detection of Cherenkov radiation in high energy physics experiments in the
USA (68) and the USSR (69). The basic concept is shown in cross-section in
Fig. 3.12 and is known as the compound parabolic concentrator. Concentration
factors of up to ten can be achieved without diurnal tracking and if lower
factors, in the order of three, are acceptable, even seasonal adjustments
may not be needed. The efficiency for accepting diffuse radiation is much
larger than for conventional focusing collectors and is the reciprocal of
the concentration factor. As shown in Fig. 3.12, the focus of the right-
hand parabola is at the base of the left-hand parabola, and vice versa.
The axis of each parabola is inclined to the vertical optic axis. Heat
collection can be achieved by adding a cylindrical black body collector at
the base of the parabolic array or by extending the parabola to encircle any
particular collector shape. A review of some of these alternatives has been
given by Rabl (70).

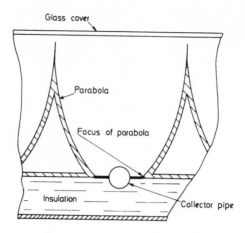

Fig. 3.12.   Concentrating parabolic collector.

Although focusing devices had been ignored in the UK because of the
relatively large amount of diffuse radiation, an analysis I carried out in
1978 indicated that the importance of the direct beam radiation component
in heating applications had been greatly underestimated (71). Tests carried
out with my double cusp-type collector shown in Fig. 3.13 showed that it
performed as well as many conventional double-glazed or selectively coated
single-glazed flat plate collectors. Practical details arising in the
design of CPC collectors were discussed in considerable detail by Rabl,
Goodman and Winston (72) in 1979. A useful design approach to the problem
of selecting the optimum depth of cylindrical parabolic concentrating
collectors has been suggested by Tabor (73).

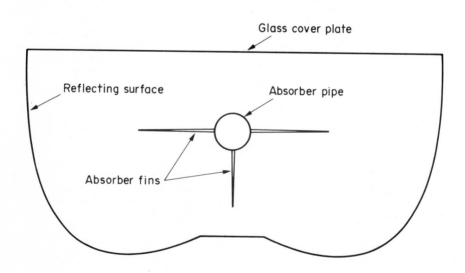

Fig . 3.13.   Double Cusp Collector.

The spiral or 'sea shell' collector. An extension of the compound parabolic concentrator into a single-sided parabolic section ending with a circular reflector was also described by Rabl (70). As shown in Fig. 3.14, a spiral collector contains an inwardly spiralling curved section. When direct radiation enters the spiral it cannot be reflected outwards and continues to be reflected deeper into the spiral until the absorbing section is reached, shown as a circular cross-section in Fig. 3.14. For the solar thermal generation of electricity Smith (74) suggests a parabolic entry section followed by a spiral curve and evacuating the space around the collector.

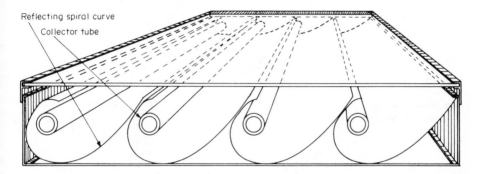

Fig. 3.14.

The trapezoidal moderately concentrating collector. Moderate concentration of solar energy can be achieved by redirecting the incident radiation on a given area onto a smaller area. Since focusing is not required, both direct and diffuse radiation can be used. A simple, easily constructed collector of this type consists of a series of long, trapezoidal, parallel stationary grooves, as shown in cross-section in Fig. 3.15. The grooves have highly reflective sidewalls and the base is the absorbing surface of the collector. Because the absorbing surface is smaller than the overall collector area, longwave emission losses are reduced. The term 'directional selectivity' can be used in describing this feature and it was demonstrated for a V-groove configuration by Hollands (75). Bannerot and Howell (76, 77) produced design nomographs for different geometries and indicated that this class of collector could have considerable potential for applications in commercially available absorption cooling, where flat plate collectors have reached the limit of their useful output at about $100-150^{\circ}C$.

An extension of this principle to the whole absorber plate has been described by Whitfield (78), who mounted the plate horizontally with a vertical mirror behind it and a horizontal mirror below. Facing south (in the northern hemisphere) it can collect about twice as much heat as a simple flat plate collector during the summer months in latitudes from 50° to 60°. The principle could probably be extended to winter use at lower latitudes with longer periods of winter sunshine.

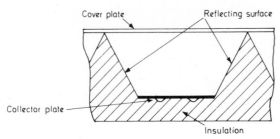

Fig. 3.15.

Evacuated Systems

An alternative approach to the problem of reducing heat losses from flat
plate collectors at high temperatures, typically in the range from 80 to
150°C, is to use an evacuated collector. The combined use of a moderate
vacuum, 1 mm Hg (133.3 N/m^2 or 1 torr), and conventional selective absorber
surfaces in Dallas, Texas (79, 80) showed that it was possible to operate at
a temperature of 150°C with a daily energy collection efficiency of more
than 40%. The spacing between the absorber surface and the glass cover was
found to be very critical in suppressing the natural convection and
conduction losses. There are various practical difficulties with this
system, but they are not considered to be formidable. Early tests
experienced trouble with seals, but a technique of using high temperature
silicone sealants has been developed. Acrylic covers used in the early tests
have been replaced by tempered or chemically-strengthened glass covers.
This system was developed into a pre-production commercial prototype in 1975
(81).

Several commercial groups have developed evacuated tubular collectors (27,
82, 83, 87) and the basic module of the Owens-Illinois collector, first
demonstrated in 1975, is illustrated in Fig. 3.16. Each module contains 24
tubes, nominally 50 mm diameter × 1.12 m long and several major commercial
systems had been installed by 1975, including a 46.5 m^2 system for cooling in
Los Angeles and a 93 m^2 system in an office building in Detroit. A cross-
section through a tube is shown in Fig. 3.17 compared with the early Philips
type. The Owens-Illinois tube has a vacuum pressure of less than 10^{-4} mm Hg,
a cover tube transmittance of 0.92 and a selective coating applied to the
outer surface of the absorber tube (τ = 0.86 and e = 0.07). The absorber
tube is supported at its free end by a spring clip to allow for differential
expansion and is hermetically sealed to the cover tube. The feeder tube
provides the reverse flow path for the heat transfer fluid. An inexpensive
diffuse reflecting surface behind the tubes has been shown to nearly double
the energy intercepted by the tube (82). The Philips collector was a double
absorber tube system with the outer cover having an internal transparent
selective coating on its upper half and a reflecting mirror surface on its
lower half. The transparent selective coating of indium oxide (In_2O_3) has a
transmittance τ of 0.85 and a reflectance ρ of about 0.9 in the range of
infra-red radiation from the absorber tubes between 300 and 400 K. The
Philips group have emphasized the good performance which can be obtained with
this system under the generally diffuse radiation conditions of northern
Europe. Confirmation of the need for a vacuum pressure less than 10^{-4} mm Hg
with this type of collector was given by work reported from Australia (84),
where moderate vacuums in the order of 0.5 mm Hg produced no improvement in
the performance. An important feature common to all tubular collectors is

that the reflection losses with direct radiation will be much lower
compared with a flat glazed surface. This enables more use to be made of
early morning and late afternoon direct radiation.

Fig. 3.16.

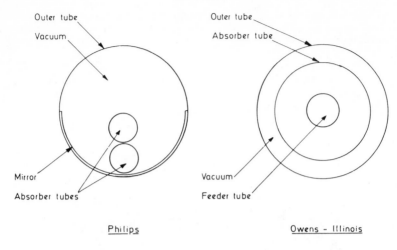

Fig. 3.17.   Evacuated tubular collectors.

By 1981, Collins (85) at the University of Sydney had developed a new
selective surface for a glass tube with outstandingly good characteristics
by using a d.c. sputtering process in low pressure gas.   A copper film is
first deposited in pure argon onto a glass tube.   This film has very low
emittance and is subsequently coated with an anti reflection layer.
Typical absorptance (α) values of 0.93 and emittance (ϵ) of 0.035 at 20°C
rising to 0.045 at 150°C were claimed for this method.   Another technique,
electroplating black-chrome onto glass tubes (α = 0.94, ϵ = 0.08) has been
described by Grimmer and Collier (86).

Several different types of evacuated tubular collector tested in various
projects in the USA and Europe up to 1981 have also been described (87),
with typical values of ($\tau\alpha$) between 0.84 and 0.86 and U 1.5 W/m^2 K.   A
significant development at this point was the combination of a selective
surface on a heat pipe collector tube inside the evacuated tube.   One system
also includes a symmetric cusp reflector fitted into a transparent glass
tube with a high reflectance, low concentration cusp mirror (88), while
other types have had both specular and diffuse reflectors (89).

The Heat Pipe Collector

The basic components of a heat pipe are shown in Fig. 3.18.   A small amount
of liquid, which is in equilibrium with its saturated vapour is sealed
within the pipe (although other configurations may be used).   When heat is
applied at one end, evaporation takes place and the excess vapour is
condensed at the other, unheated, end of the heat pipe.   The condensate is
returned to the heated end by means of capillary forces in the wick section.
In some solar heating applications, the return of the condensate can be
simply achieved by gravity flow.   As the process of evaporation and
condensation at constant pressure is also a constant temperature operation,
the heat pipe is capable of transferring thermal energy internally with very
small temperature differences.   There is an inevitable loss of efficiency in
transferring the heat from the heat pipe to the secondary circuit.   A major
programme of heat pipe collection research has been carried out in the USA

since 1974 (90) and work in Holland during 1975 was reported by Francken
(91), who emphasized the fast thermal response characteristics to changes in
solar radiation.   Another advantage is that it can contain fluids with lower
freezing points than water.

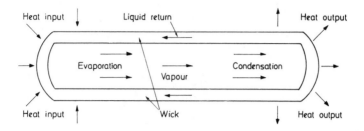

Fig. 3.18.   Heat pipe collector.

By the early 1980s heat pipe collectors had become well established in the
commercial market and their performance was substantially better than the
simpler 'first generation' flat plate collectors seen some ten years
earlier, with instantaneous efficiencies of 40% and above under conditions
where the simple single-glazed collector would have ceased to operate.
This is shown in Fig. 3.24.

The Floating-deck Heater

The philosophy behind this development (92) is that considering the diffuse
nature of solar energy, there is as much to be gained by simplicity, low
cost and ease of installation as in increased efficiency, although the
experimental evidence showed that its efficiency was comparable with other
horizontal flat plate collectors.   The basic principles are shown in
Fig. 3.19.   The floating deck is an insulating layer, preferably foam glass,
which floats on a hot water storage section.   Solar energy is collected by
the water flowing in a thin film over the top of the insulation.   A
compressing layer, which can be transparent or black glass, plastic or metal,
rests directly upon the floating water film.   After initial tests on a square
heater with an area of 0.836 m^2, a large-scale version with an area of
46.5 m^2 has been successfully developed.   It was considered to be suitable
for fairly low temperature applications in the lower latitudes, or possibly
in combination with a long term storage facility at higher latitudes.

The Cylindrical Heater/Storage System

This is a self-contained cylindrical heat collector and storage vessel
developed over several years in New Zealand by Vincze (93, 94).   The
operating principles are illustrated in Fig. 3.20.   As the solar radiation
reaches the black collector surface, the water in the narrow annular space
rises and the cooler water inside the vessel descends, establishing a natural
thermosyphon.   Test results indicated (94) that it has a superior performance
when compared with a flat plate collector, provided the actual area of the
projected cylinder is used as the basis for comparison.   If the area needed
to space the cylinders apart is taken into account, then the flat plate and
the cylindrical have very similar efficiencies.

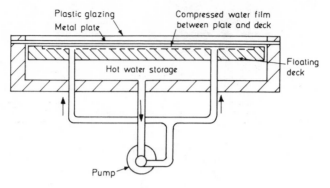

Fig. 3.19.   The floating-deck heater.

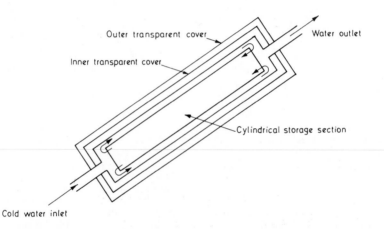

Fig. 3.20.   The cylindrical heater/storage system.

The Thermic Diode Panel

The thermic diode panel, developed at the Massachusetts Institute of
Technology by Professor Shawn Buckley (95-97) allows heat to flow in only
one direction, thus preventing heat losses from a building at night or
during cloudy days. The panel is unique in that it contains all the
elements of a complete solar energy system: collectors, controls, storage,
heat exchangers and ducting. The basic operation is shown in Fig. 3.21 and
is similar to a thermosyphon solar system except that the storage need not
be located above the collector. During the heating operation the check
valve is open. This valve is the key to the system and consists of a simple
layer of oil floating on water. When the collector panel has ceased to
receive solar radiation and is at a lower temperature than the storage
section, the resulting reverse pressure pushes some oil down the riser tube
until the pressure is balanced by the buoyancy of the oil. This completely
checks the backward flow of the heated stored water into the cooler
collector panel. A domestic water heating version was planned for

production in the USA early in 1982 (97).

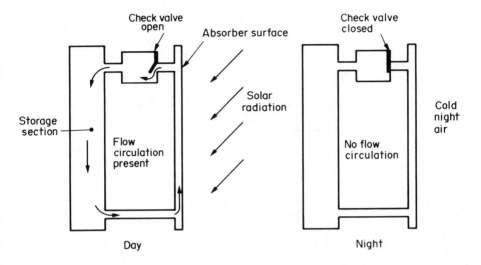

Fig. 3.21.   Basic operation of the thermic diode.

AIR HEATERS

Solar air heaters have not attracted nearly as much research and development work as water heating systems (7, 98), but there are many applications where air is a more appropriate heat transfer fluid, for example crop drying in the lower latitudes or space heating in the higher latitudes. Air heaters have three particular advantages:

(i)   air cannot freeze;
(ii)   while air can leak, it is not nearly so serious as water leakage;
(iii) problems of corrosion in mixed metal systems and storage tanks are less likely.

However, the physical properties of air are less favourable, particularly its relatively very low density and specific heat, and the ducts needed in air systems are very much larger than water pipes.

Simple air heaters can be made from almost any surface which can be painted black. The three main types of simple heater are shown in Fig. 3.22, with single cover plates. In type (a), the duct is the space between the transparent cover and the collector absorber plate. With type (b), there is a sealed air gap between the cover plate and the collector plate to reduce convective heat exchange, and the duct is behind the collector plate. Type (c) uses either a divided air stream or takes the air flow through the outer section as a preheater before passing through the inner air duct. An excellent example of a solar air heater made out of simple materials is an installation with over 500 m^2 of collector in Gujarat (99), where the air is passed through black painted swarf, the metal scrap discarded after metal cutting processes. The double-glazed collector has an estimated efficiency of about 45% with a temperature difference of 65°C above ambient. The early work by Lof in the USA (100) was carried out with a series of overlapping

black glass plates installed under one, two or three cover plates. Lof
subsequently installed an overlapped glass plate system in the Colorado
Solar House and its performance in the 1959-1960 heating season has been
reported (101). After 16 years of practically trouble free operation, the
system was included in the 1976-1977 US programme (102). Australia has been
another centre for major solar air heater research and the use of selective
surfaces in V-corrugated surfaces was first studied there in the early 1960s
(75, 103). The air heater shown in Fig. 3.2(i) is based on this principle.

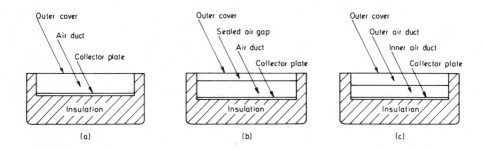

Fig. 3.22. Air heaters.

As well as selective surfaces, methods of improving collector efficiency
include control of the air velocity and the use of a two-pass system (104),
in which the air passes between the two glass cover plates of an otherwise
conventional double-glazed unit. This particular system showed an effective
increase in efficiency of up to 17% compared with operation as a conventional
unit. Other systems use various types of finned surface to improve heat
transfer (105) as shown in Fig. 3.2(h), and the use of honeycombs combined
with a porous bed has also been tried (106).

The integration of an air collector into a heating and cooling system is
shown in Fig. 3.23, which is based on diagrams given by Löf (98). The
storage unit is of the 'rock storage' type and in this case consists of
ordinary screened gravel. The fan and control unit can direct the air flow
in any of the following modes:

  (i)   direct heating of the house from the collector;
  (ii) house heating from the storage unit;
  (iii) storage of heat from the collector;
  (iv) cooling the storage unit from external cold air;
  (v)   house cooling from the storage unit.

The dual use of the storage unit for both summer cooling and winter heating
is an added attraction. The auxiliary heating system is omitted from the
diagram.

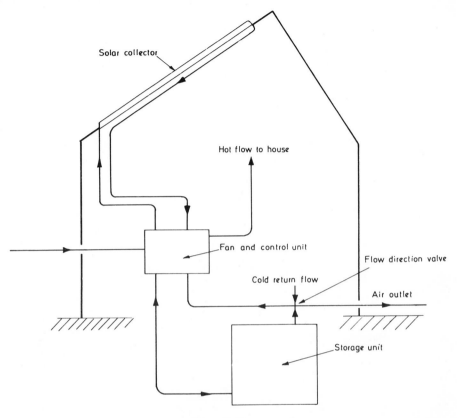

Fig. 3.23.

An interesting innovation in vertical air collector design was described by
Hummel, Markowitz and Lee (107) in 1981. Convective heat losses were
suppressed by directing and controlling a downward flowing cold boundary
layer just inside the glazing while an inner upward flowing layer is heated
as it passes over and through a series of selectively coated absorber plates.
Early results showed an efficiency intercept of 0.81 and a slope of
2.9 W/m^2 $^\circ$C, which indicated a substantial improvement over the earlier
designs of Löf and Heywood, as shown in Fig. 3.24.

COMPARATIVE PERFORMANCE OF COLLECTORS

Figure 3.24 shows the efficiency plotted against $(T_m - T_a) \times G_c^{-1}$ for 14
different types of collector, based on data available in 1976, compared with
some 1981 results.

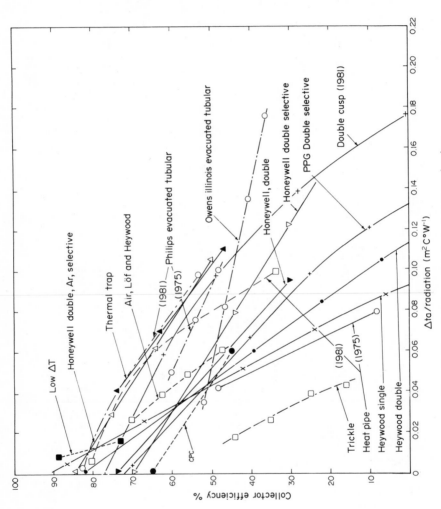

Fig. 3.24. Solar collector characteristics.

The characteristic curves of all but the two simplest types, the low
temperature rise (48) and the trickle (55, 108), pass through the rectangle
bounded by 50-70% efficiency and 0.03-0.05 $(T_m - T_a) \times G_c^{-1}$. This means
that for incident radiation intensities greater than 500 W/m^2 all these
heaters will have similar performances for temperature rises between 15 and
30°C above ambient, the most common range for domestic water heating
applications. The trickle type and the low temperature rise type are not
suitable for high temperature applications, as the maximum possible rise
appears to be about 60°C at zero efficiency. A commercial flat plate
collector with a good performance characteristic, the Honeywell double-glazed
collector with an anti-reflection (AR) surface on the glazing and a selective
surface on the collector absorber plate (109), has a very similar performance
to the conceptually simpler thermal trap system (55) but the latter had not
attracted commercial support by the early 1980s. Heat pipe collectors have
improved significantly, as the comparative performance of the 1976 and 1981
versions show. Double-glazed cusp collectors and collectors using heat pipe
principles with various refrigerants as the working fluid all showed
performances in the same region during 1981. Data from tests on a
fluorocarbon (trichlorofluoromethane) collector system show high average
efficiencies of 83% at low values of ΔT (110).

Preliminary results from two 1976 evacuated tubular collectors (27, 82)
showed that they performed very well at high temperature differentials under
good radiation conditions, but also had a good performance under poor
radiation conditions. The performance of the best 1981 Philips evacuated
tubular collector shows a significant improvement compared with the earlier
collectors. Heywood's early results for double and single glazing (3) have
long been used as a standard for all simple flat plate collectors and it was
interesting to see that one major commercial organization had exhibited a
'new' collector at the 1981 Solar World Forum with an inferior performance to
Heywood's in the 1950s. The PPG collector (111) is representative of a more
advanced commercially available type. Simple air heaters (3, 101) have a
relatively good performance when compared with conventional water heaters.

Although Fig. 3.24 is based on an equation with many simplifying assumptions,
it enables a fair comparison to be made between collectors tested in
different locations under very different radiation conditions. This type of
graph has become increasingly important as national standards of performance
have become established throughout the world. However, it does not give an
economic assessment or comparison. Collectors with very similar thermal
performance characteristics could differ by a factor of at least two in their
cost. The estimated service life is another important practical feature
which cannot be evaluated in this analysis.

TEST PROCEDURES

With an increasing number of new solar heaters appearing in the late 1970s,
it was important that an internationally agreed standard testing procedure
could be established. The first country to establish a national standard
was Israel in 1966 (112). This followed earlier work at the National
Physical Laboratory of Israel (113) and Tabor (114) who outlined a testing
procedure using a governing equation which is essentially equation (3.2),
the modified Hottel-Whillier-Bliss equation. Tabor's suggested experimental
procedure involves the connection of up to four similar collectors in series.
At any instant of time the solar radiation intensity and flow rates are
identical for each collector and a number of points on the performance
characteristic curve, similar to Fig. 3.24, can be obtained from one test.

The procedure also calls for testing on a clear day with little wind and estimates for a typical flat plate collector show the efficiency for values of $(T_m - T_a)$ = 4.5°C to drop about 0.5% for a wind change from zero to 4.47 m/s, 6.5% at $(T_m - T_a)$ = 26.7°C and 19.5% at $(T_m - T_a)$ = 48.9°C. A non-linear theoretical model to allow for the effect of wind velocity has been suggested in Australia (115).

By 1975 draft standards for both solar collectors (116) and thermal storage devices (117) had been published in the USA. These also used the concept of equation (3.2) and defined in considerable detail the methods of measuring the various parameters, such as temperature, pressure, flow rate and radiation intensity. At least four test points at recommended values of 10, 30, 50 and 70°C for $(T_m - T_a)$ were suggested to establish the characteristic curve of any collector.

A major assessment of solar heating standards activities in the USA by 1980 was given by Waksman and Dikkers (118). The paper identified the major US public and private sector organizations involved in the development and implementation of standards for solar heating applications and included 48 references. Many of the references were to new solar standards issued by the National Bureau of Standards. Brinkworth (119) has described the development of British Standards for solar heating and extracts from BS 5918: 1980, Code of Practice for Solar Heating Systems for Domestic Hot Water, are included in Chapter 10. Solar collector testing activities in the EEC have resulted in the establishment of a working group of 20 laboratories (120) and work on model validation for a single family dwelling in Europe, with a Solar Pilot Test Facility within each participating country, has been described by La Fontaine (121).

The use of a solar simulator, or artificial sun, enables a collector to be tested under standard conditions of ambient temperature, wind and radiation intensity and performance results obtained with a simulator in the USA (122, 123) have been found to be in good agreement with outdoor performance results. A solar simulator facility forms part of the research programme at the University of Cardiff in the UK and was described by Gillet in 1980 (124). Subsequently Gillet showed that variations in collector efficiency between indoor and outdoor tests are small for 'good' collectors, but that in the case of poor collector designs a mixed indoor/outdoor test will overestimate the collection efficiency. In countries with variable radiation conditions, solar simulators can play an important part in testing procedures. Among the more sophisticated testing procedures beginning to be more widely known in the 1980s, the use of infra-red thermography was reported from several countries including the USA (125) and the UK (126). This technique enables an observer to 'see' the temperature differences on the surface of a collector as different bands of colour and will be a valuable diagnostic tool when it has been developed.

Details of field trials on domestic solar water systems installed in dwellings carried out by the Building Research Establishment (127) have been published and indicate that there are still many areas of uncertainty. For example, one objective for the 1982 tests was to try to find out why nominally identical solar systems serving households having the same average daily use of hot water have different observed efficiencies.

REFERENCES

(1) Brooks, F. A., Solar energy and its uses for heating water in
     California, Bull. Calif. Agric. Exp. Sta., No. 602, 1936.

(2) Hottel, H. C. and Woertz, B. B., The performance of flat plate solar
     heat collectors, *Trans. ASME*, 64, 91-104, 1942.

(3) Heywood, H., Solar energy for water and space heating, *J. Inst. Fuel*
     27, 334-347, July 1954.

(4) Morse, R. N., Solar water heaters, Proc. World Symposium on Applied
     Solar Energy, Stanford Research Inst., University of Arizona, Phoenix,
     Arizona, 191-202, 1956.

(5) Chinnery, D. N. W., Solar water heating in South Africa, National
     Building Research Institute, Bulletin 44, CSIR Research Report 248,
     Pretoria, South Africa, 1967.

(6) Tabor, H., Solar energy collector design, Bull. Res. Coun., 5C, No. 1,
     Israel, 1955.

(7) Yellott, J. I., Solar energy utilization for heating and cooling,
     originally published in ASHRAE Journal, December 1973, now in
     Chapter 59, ASHRAE Guide and Data Book series, 1974 edition.

(8) Scott, J. E., User's experience with solar water heater collectors in
     Florida, Proc. Workshop on Solar Collectors for Heating and Cooling of
     Buildings, 21-23 November 1974, NSF-RANN-75-019, May 1975.

(9) CSIRO Solar Water Heaters, Division of Mechanical Engineering Circular
     No. 2, Melbourne, 1964.

(10) McVeigh, J. C., Success or Failure? Some factors influencing the
     development of the solar water heating industry in the United Kingdom
     UK ISES Conference on Practical Applications of Solar Energy in
     Industry and Commerce, pp. 104-108, Birmingham, Midlands Branch, 1979.

(11) Cross, B. M. and Johansson, C. M. A., Solar Collector and performance
     in the UK, Solar World Forum, Vol.2, 1526-1530, Pergamon Press,
     Oxford, 1982.

(12) New and Renewable Energy in the Arab World, United Nations Economic
     Commission for Western Asia, Beirut, 1981.

(13) McVeigh, J. C., Some experiments with a flat plate solar water heater,
     UK Section, ISES, Conf. on Low Temperature Thermal Collection of Solar
     Energy, April 1974.

(14) McVeigh, J. C., Low-cost solar water heater, Proc. Conf. on Appropriate
     Technology, University of Newcastle upon Tyne, 1976.

(15) Whillier, A., Solar energy collection and its utilization for house
     heating, ScD. thesis, MIT, 1953.

(16) Whillier, A., Design factors influencing collector performance, Low
     Temperature Engineering Applications of Solar Energy, ASHRAE, New York,
     1967.

(17) Hottel, H. C. and Whillier, A., Evaluation of flat plate collector performance, Trans. Conf. on the Use of Solar Energy, 2(1), 74, University of Arizona Press, 1958.

(18) Bliss, R. W., The derivation of several 'plate efficiency factors', useful in the design of flat plate solar heat collectors, *Solar Energy* 3, (4), 55, 1959.

(19) Duffie, J. A. and Beckman, W. A., *Solar Energy Thermal Processes*, John Wiley, New York, 1974.

(20) Smith, C. T. and Weiss, T. A., Design applications of the Hottel-Whillier-Bliss equation, ISES Congress, Los Angeles, Extended Abstracts, Paper 34/6, July 1975, and in *Solar Energy* 19, 109-113, 1977.

(21) Mitalas, G. P. and Stephenson, D. G., Absorption and Transmission of Thermal Radiation by Single and Double Glazed Windows, Research paper 173, Division of Building Research, National Research Council of Canada, Ottawa, 1962.

(22) Zarem, A. M. and Erway, D. D., *Introduction to the Utilization of Solar Energy*, McGraw-Hill, New York, 1963.

(23) Scoville, A. E., An alternate cover material for solar collectors, ISES Congress, Los Angeles, Extended Abstracts, Paper 30/11, July 1975.

(24) Charters, W. W. S. and Macdonald, R. W. G., Heat transfer effects in solar air heaters, Paper E 37. Conf. The Sun in the Service of Mankind, UNESCO, Paris, 1973.

(25) Lorsch, H. G., Performance of flat plate collectors, Proc. Solar Heating and Cooling for Buildings Workshop, 21-23 March 1973, NSF-RANN-73-004, July 1973.

(26) Tani, T., Sawata, S., Tanaka, T. and Horigome, T., Characteristics of selective thin barriers and selective surfaces, ISES Congress, Los Angeles, Extended Abstracts, Paper 30/4, July 1975, and in *Solar Energy* 18, 281-285, 1976.

(27) Bruno, R., Herman, W., Horster, K., Kersten, R. and Mahdjuri, R., High efficiency solar collectors, Paper 34/10, *ibid.*

(28) Löf, G. O. G. and French, R. R., Hail resistance of solar collectors with tempered glass covers, *Solar Energy* 25, 555-561, 1980.

(29) Cox, M. and Armstrong, P. R., A statistical Model for the Incidence of Large Hailstones on Solar Collectors, *Solar Energy* 26, 97-111, 1981.

(30) McDonald, G. E., Spectral reflectance properties of plate zinc for use as a solar selective coating, ISES Congress, Los Angeles, Extended Abstracts Paper 30/2, July 1975.

(31) Solar Energy: a UK Assessment, UK Section, ISES, London, May 1976.

(32) Selective Black Coatings, Proc. UN Conf. on New Sources of Energy, 4, 618, 1964.

(33) Close, D. J., Flat plate solar absorbers. The production and testing
      of a selective surface for copper absorber plates. Report ED 7, CSIRO,
      Melbourne, June 1962.

(34) Keller, A., Selective surfaces of copper foils, Paper E 43, Conf. The
      Sun in the Service of Mankind, UNESCO, Paris, 1973.

(35) Keller, A., Selective surfaces of aluminium foils, Paper 7/16,
      International Solar Energy Society Conference, Melbourne, 1970.

(36) McDonald, G. E., Refinement in Black Chrome for use as a solar
      selective coating, NASA TM X-3136, 1975.

(37) Pettit, R. B. and Sowell, R. R., Solar absorptance and emittance
      properties of several solar coatings, ISES Congress, Los Angeles,
      Extended Abstracts, Paper 30/1, July 1975.

(38) Harris, L., The optical properties of metal blacks and carbon blacks,
      The Eppley Foundation for Research Monograph Series No. 1, 1967.

(39) McKenzie, D. R., Harding, G. L. and Window, B., Metal blacks as
      selective surfaces, ISES Congress, Los Angeles, Extended Abstracts,
      Paper 30/5, July 1975.

(40) Sabine, T. M., Gammon, R. B. and Riddiford, C. L., 'Solarox' as a
      selective absorber, Paper 30/6, *ibid.*

(41) Mason, J. J. and Jones, P. C., The use of selective foils for coating
      and fabricating solar absorbers, Solar World Forum, Vol. 1, 237-242,
      Pergamon Press, Oxford, 1982.

(42) Stamford, M. S., Compatibility of solar systems, *A question of solar
      heating,* Copper Development Association, Potters Bar, Hertfordshire,
      1976.

(43) Private communication, Department of Mechanical Engineering and Energy
      Studies, University of Cardiff, 1976.

(44) Duffie, J. A. *et al.,* Report of working group on materials and
      components for flat plate collectors, Proc. Workshop on Solar
      Collectors for Heating and Cooling of Buildings, 21-23 November, 1974,
      NSF-RANN-75-019, May 1975.

(45) Popplewell, J. M., Corrosion considerations in the use of aluminium and
      copper solar energy collectors, ISES Congress, Los Angeles, Extended
      Abstracts, Paper 30/12, July 1975.

(46) Olsson, G., Thundal, B. and Wilson, G., Advanced solar absorber of
      metallurgically bonded copper and aluminium, Solar World Forum, Vol. 1,
      163-167, Pergamon Press, Oxford, 1982.

(47) Justin, B., Some aspects of testing the thermal performance of
      collectors, UK ISES Conference (C11) on Testing of Solar Collectors and
      Systems, 1977.

(48) McVeigh, J. C., Some experiments in heating swimming pools by solar
      energy, *JIHVE* 39, 53-55, June 1971.

(49)   Czarnecki, J. T., Method of heating swimming pools by solar energy, *Solar Energy* 7, 3-7, 1963.

(50)   Root, D. E., Practical aspects of swimming pool heating, *Solar Energy* 4(1), 23-24, 1960.

(51)   Farber, E. A. and Triandafyllis, J., Solar swimming pool heating, Conf. The Sun in the Service of Mankind, UNESCO, Paris, 1973.

(52)   Farber, E. A., Solar energy research and development at the University of Florida, Building Systems Design, February/March, 1974.

(53)   UK ISES Conference Paper (C17) on Practical experiences with solar heated swimming pools, October 1978.

(54)   Cobble, M. H., Irradiation into transparent solids and the thermal trap effect, *J. Frank. Inst.* 278, 383-393, 1964.

(55)   San Martin, R. L. and Fjeld, G. J., Experimental performance of three solar collectors, *Solar Energy* 17, 345-349, 1975.

(56)   Samuel, T. D. M. A. and Wijeysundera, N. E., Optimum performance of thermal trap collectors, *Solar Energy* 26, 65-76, 1981.

(57)   Marshall, K. N., Bell, G. A., Wedel, R. K. and Haslim, L. A., Thermal radiation characteristics of transparent plastic honeycombs for solar collector applications, ISES Congress, Los Angeles, Extended Abstracts, Paper 32/1, July 1975.

(58)   Cane, R. L. D., Hollands, K. G. T., Raithby, G. D. and Unny, T. E., Convection suppression in inclined honeycombs, Paper 32/5, *Ibid.*

(59)   Buchberg, H., Edwards, D. K. and Mackenzie, J. D., Design considerations for solar collectors with glass cylindrical cellular covers, Paper 32/12, *Ibid.* and *Solar Energy,* 18, 193-203, 1976.

(60)   Baldwin, C. M., Dunn, B. S., Hilliard, W. G. and Mackenzie, J. D., Performance of transparent glass honeycombs in flat plate collectors, Paper 32/2, *Ibid.*

(61)   Edwards, D. K., Arnold, J. N. and Catton, I., End clearance effects on rectangular-honeycomb solar collectors, *Solar Energy,* 18, 253-257, 1976.

(62)   Wu, P. S. and Edwards, D. K., Effect of combined tilt and end clearance upon natural convection in high L/D rectangular honeycomb, *Solar Energy* 25, 471-473, 1980.

(63)   Moore, S. W., Balcomb, J. D. and Hedstrom, J. C., Design and testing of a structurally integrated steel solar collector unit based on expanded flat metal plates, Presented at US Section ISES Meeting, Fort Collins, Colorado, 19-23 August 1974.

(64)   Balcomb, J. D., Hedstrom, J. C. and Moore, S. W., The LASL structurally integrated solar collector unit - final results, ISES Congress, Los Angeles, Extended Abstracts, Paper 34/1, July 1975.

(65) Spencer, D. L., Smith, T. F. and Flindt, H. R., The design and
      performance of a distributed flow water-cooled solar collector,
      College of Engineering, University of Iowa, Iowa 52242, 1975.

(66) Smith, T. F., Jensen, P. A. and Spencer, D. L., Thermal performance of
      the distributed flow, subatmospheric pressure, flat plate solar
      collector, *Solar Energy* 25, 429-436, 1980.

(67) Winston, R., Principles of solar concentrators of a novel design,
      Solar Energy 16, 89-95, 1974.

(68) Hinterberger, H. and Winston, R., *Rev. Sci. Instr.* 37, 1094, 1966.

(69) Varanov, V. K. and Melnikov, G. K., *Soviet Journal of Optical
      Technology* 33, 408, 1966.

(70) Rabl. A., Comparison of solar concentrators, *Solar Energy* 18, 93-111,
      1976.

(71) McVeigh, J. C., A stationary concentrator for industry, UK ISES
      Conference (C14) on Solar Energy for Industry, London, February 1978.

(72) Rabl, A. and Goodman, N. B. and Winston, R., Practical design
      considerations for CPC solar collectors, *Solar Energy* 22, 373-381,
      1979.

(73) Tabor, H., A note of the economics of deep cylindrical mirror
      concentrating collectors, *Solar Energy* 19, 573-574, 1977.

(74) Smith, R. H., A method of solar thermal generation of electricity,
      ISES Congress, Los Angeles, Extended Abstracts, Paper 53/8, July 1975.

(75) Hollands, K. G. T., Directional selectivity, emittance and absorptance
      properties of vee corrugated specular surfaces, *Solar Energy* 7,
      108-116, 1963.

(76) Bannerot, R. B. and Howell, J. R., Moderately concentrating flat plate
      solar energy collectors, ASME paper 75-HT-54, presented at AIChE-ASME
      Heat Transfer Conference, San Francisco, California, 11-13 August
      1975.

(77) Bannerot, R. B. and Howell, J. R., The effect of non-direct insolation
      on the radiative performance of trapezoidal grooves used as solar
      energy collectors, ISES Congress, Los Angeles, Extended Abstracts.
      Paper 52/5, July 1975, and in *Solar Energy* 19, 549-553, 1977.

(78) Whitfield, G. R., The plane mirror concentrating flat plate collector,
      Solar World Forum, Vol. 1, 141-145, Pergamon Press, Oxford 1982.

(79) Blum, H. A. and Estes, J. M., Design and feasibility of flat plate
      solar collectors to operate at 100-150°C, Paper E 18, Conf. The Sun in
      the Service of Mankind, UNESCO, Paris, 1973.

(80) Eaton, C. B. and Blum, H. A., The use of moderate vacuum environments
      as a means of increasing the collection efficiencies and operating
      temperatures of flat plate solar collectors, *Solar Energy* 17, 151-158,
      1975.

(81) Kittle, P. A. and Cope, S. L., Outside performance of moderate vacuum
      solar collectors, ISES Congress, Los Angeles, Extended Abstracts,
      Paper 32/8, July 1975.

(82) Beekley, D. C. and Mather, J. R., Analysis and experimental tests of
      high performance tubular solar collectors, Paper 32/10, *Ibid.*

(83) Ortabasi, U. and Buehl, W. M., Analysis and performance of an
      evacuated tubular collector, Paper 32/11, *Ibid.*

(84) Read, W. R. and Christie, E. A., Thermal characteristics of evacuated
      tubular solar collectors, Paper 32/9, *Ibid.*

(85) Collins, R. E., High temperature solar collectors, Solar Energy Group,
      University of Sydney, Sydney, NSW 2006, Australia, 1980.

(86) Grimmer, D. P. and Collier, R. K., Black-chrome solar-selective
      coatings electrodeposited on metallized glass tubes, *Solar Energy* 26,
      467-469, 1981.

(87) Hörster, H. and Kersten, R., Evacuated Solar Collectors, UK ISES
      Conference (C25) on Recent Developments in Solar Collector Design,
      London, January 1981.

(88) Ortabasi, U. and Buehl, W. M., An internal cusp reflector for an
      evacuated tubular heat pipe solar thermal collector, *Solar Energy* 25,
      67-78, 1980.

(89) Bloem, H., de Grijs, J. C. and de Vaan, R. L. C., Evacuated tubular
      collector with two-phase heat transfer into the system, Solar World
      Forum, Vol. 1, 176-180, Pergamon Press, Oxford, 1982.

(90) Bienert, W. B., Heat pipes applied to flat plate solar collectors,
      Proc. Workshop on Solar Collectors for Heating and Cooling of Buildings,
      21-23 November 1974, NSF-RANN-75-019, May 1975.

(91) Francken, J. C., The heat pipe fin, a novel design of a planar
      collector, ISES Congress, Los Angeles, Extended Abstracts, Paper 34/8,
      July 1975.

(92) Davison, R. R., Harris, W. B. and Chan Ho Kai, Design and performance
      of the compressed-film floating deck solar water heater, ISES Congress,
      Los Angeles, Extended Abstracts, Paper 34/9, July 1975.

(93) Vincze, S. A., A high-speed cylindrical solar water heater, *Solar
      Energy* 13, 339-344 1971.

(94) Vincze, S. A., Comparative winter tests, cylindrical versus flat plate
      solar heat collectors, ISES Congress, Los Angeles, Extended Abstracts,
      Paper 14/8, July 1975.

(95) Buckley, S., The thermic diode solar panel, Sunworld, pp. 7-9, August
      1977.

(96) Buckley, S., Thermic diode solar panels for space heating, *Solar
      Energy,* 20, 495-503, 1978.

(97) Buckley, S., Reply to C. Mustacchi *et al.*, *Solar Energy,* 25, 577-578,
      1980.

(98) Löf, G. O. G., Space heating with solar air collectors, Proc. Workshop on Solar Collectors for Heating and Cooling of Buildings, 21-23 November 1974, NSF-RANN-75-019, May 1975.

(99) Chandran, T. C., Private Communication, Kaira District Co-operative Milk Producers' Union Ltd., Anand. 388001, Gujarat, India, 1976.

(100) Löf, G. O. G. and Nevens, T. D., Heating of air by solar energy, *Ohio Journal of Science* 53, 272-280, 1953.

(101) Löf, G. O. G., El Wakil, M. M. and Chion, J. P., Design and performance of domestic heating system employing solar-heated air - the Colorado Solar House, Proc. UN Conf. New Sources of Energy, 185-197, 1964.

(102) Ward, J. C., Long term (16 years) performance of an overlapped-glass plate solar-air heater, Proc. Workshop on Solar Collectors for Heating and Cooling of Buildings, 21-23 November 1974, NSF-RANN-75-019, May 1975.

(103) Close, D. J., Solar air heaters for low and moderate temperature application, *Solar Energy* 7, 117-124, 1963.

(104) Satcunanathan, S. and Deonarine, S., A two-pass solar air heater, *Solar Energy* 15, 41-49, 1973.

(105) Bevill, V. D. and Brandt, H., A solar energy collector for heating air, *Solar Energy* 12, 19-29, 1968.

(106) Lalude, O. A. and Buchberg, H., Design and application of honeycomb porous-bed solar air heaters, *Solar Energy* 13, 223-242, 1971.

(107) Hummel, R. L., Markowitz, T. and Lee, D., A High Performance, Cost Effective Solar Collector, Solar World Forum, Vol.1., 197, Pergamon Press, Oxford, 1982.

(108) Brachi, P., Sun on the roof, *New Scientist*, 19 September 1974.

(109) Ramsey, J. W. and Borzoni, J. T., Effects of selective coatings on flat plate solar collector performance, ISES Congress, Los Angeles, Extended Abstracts, Paper 34/5, July 1975.

(110) Schreyer, J. M., Residential application of refrigerant-charged solar collectors, *Solar Energy* 26, 307-312, 1981.

(111) PPG Industries, Baseline Solar Collector, One Gateway Center, Pittsburgh, PA 15222, USA, 1974.

(112) Israeli Standard No. 609, Solar Water Heaters: Test Methods, Israeli Standards Institute, Tel Aviv, May 1966.

(113) Doron, B., Testing of solar collectors, *Solar Energy* 9, 103-104, 1965.

(114) Tabor, H., The testing of solar collectors, The Scientific Research Foundation, Jerusalem, March 1975 and ISES Congress, Los Angeles, Extended Abstracts, Paper 33/8, July 1975, and in *Solar Energy* 20, 293-303, 1978.

(115)   Dunkle, R. V. and Cooper, P. I., A proposed method for the evaluation
        of performance parameters of flat plate solar collectors, ISES
        Congress, Los Angeles, Extended Abstracts, Paper 33/2, 1975.

(116)   Hill, J. E. and Kusada, T., Method of testing for rating solar
        collectors based on thermal performance, NBSIR 74-635, National
        Bureau of Standards, Washington, DC 20234, December 1974.

(117)   Kelly, G. E. and Hill, J. E., Method of testing for rating thermal
        storage devices based on thermal performance, NBSIR 74-634, National
        Bureau of Standards, Washington, DC 20234, May 1975.

(118)   Waksman, D. and Dikkers, R. D., Solar Heating Standards Activities
        in the United States, UK ISES Conference (C22) on Solar Energy Codes
        of Practice and Test Procedures, London, April, 1980.

(119)   Brinkworth, B. J., British Standards for Solar Heating, UK ISES
        Conference (C22) on Solar Energy Codes of Practice and Test
        Procedures, London, April 1980.

(120)   Aranovitch, E. and Roumengous, C., Solar Collector Testing Activities
        in the European Community, UK ISES Conference (C22) on Solar Energy
        Codes of Practice and Test Procedures, London, April, 1980.

(121)   Curtis, D. M. and La Fontaine, S. G. T., Model validation - its
        purpose, value and results, Solar World Forum, Vol. 3, 2659-2663,
        Pergamon Press, Oxford, 1982.

(122)   Simon, F. F. and Harlament, P., Flat plate collector performance
        evaluation: the case for a solar simulator approach, NASA TM X-71427,
        October 1973.

(123)   Simon, F. F., Flat plate solar collector performance evaluation with
        a solar simulator as a basis for collector selection and performance
        prediction, ISES Congress, Los Angeles, Extended Abstracts, Paper
        33/4, July 1975, and in *Solar Energy* 18, 451-466, 1976.

(124)   Gillett, W. B., The Equivalence of outdoor and mixed indoor/outdoor
        solar collector testing, *Solar Energy* 25, 543-548, 1980.

(125)   Mansfield, R. G. and Eden, A., The application of thermography to
        large arrays of solar energy collectors, *Solar Energy* 21, 533-537,
        1978.

(126)   Jesch, L. F., Elani, U. and Naybours, T. E., Diagnostic tools for
        flat plate solar collectors, UK ISES Conference (C25) on Recent
        developments in solar collector design, London, January 1981.

(127)   Wozniak, S. J., Performance and use of solar water heating systems,
        Solar World Forum, Vol. 1, 58-62, Pergamon Press, Oxford, 1982.

CHAPTER 4

SPACE HEATING APPLICATIONS
HISTORY AND ACTIVE SYSTEMS

The possibility of at least partially heating a building by solar energy has
been demonstrated on many occasions over the past forty years. The criteria
originally suggested by Telkes (1) in 1949 for solving the problems of
collection, storage and distribution of solar energy have been somewhat
modified since then as experience has been gained with an increasing number
of installations. In the early work, emphasis was placed on the collection
of winter sunshine. This has been broadened to include the very substantial
contribution that can be obtained from diffuse radiation. The three terms
'efficient', 'economic' and 'simple' used by Telkes in relation to the
collector have always been the goal of sound engineering practice and an
analysis of some of the earlier solar installations indicated that very few
can satisfy all three.

The problem of storing the solar energy collected in the summer for use the
following winter has attracted a tremendous research effort. Although the
principle of using a very large, heavily insulated water storage tank buried
underneath the building was described by Hottel and Woertz in 1942 (2),
their comment that the unit was known to be highly uneconomical had a
considerable influence on the direction of storage system research over the
next two decades. The effect of latitude and local radiation characteristics
are now more widely appreciated. It was originally thought that only a few
days' storage could be economically viable so that the solar energy received
during clear winter days would be made available for successive overcast
periods and that this could only occur to any extent in parts of the world
where there are appreciable amounts of winter sunshine. However, longer
storage periods of up to several months were achieved in several solar houses
during the mid-1970s and by the end of the decade interseasonal storage had
been shown to be not only possible but also cost effective. For shorter
term storage considerable reductions in the total volume of the store are
possible by the use of chemical storage methods pioneered by Telkes. She
also drew attention to the need to avoid overheating, especially in the
rapidly changing weather of spring and autumn, ensuring that the solar
heating system is definitely not heating the house in the summer.

The term 'solar house' first became familiar in the USA during the 1930s, when architects began to use large, south-facing windows to let the lower slanting rays of the winter sun penetrate into the back of the room (3). At that time there was no real distinction made between the active systems (using pumps or fans to circulate heat) and passive solar houses. It was noticed, however, that while fuel was saved during the day it was not possible to store the solar energy and that at night and during cloudy days the heat loss was so great that there was a relatively small saving on fuel during the entire heating season. Two identical test houses were built at Purdue University, Lafayette under the direction of Professor F. W. Hutchinson to obtain quantitative evidence of the fuel savings from solar gain (4-6). Both houses were fitted with sealed double-glazing, the orthodox house had a conventional window area while the solar house had a greater area of glass on the south side. Both houses were heated by electricity and thermostatically maintained at identical temperatures. The most surprising result to emerge from those tests was that the solar house required about 16% more heat than the orthodox house during the December-January test period. It was obvious that the larger solar windows dissipated a great deal of heat at night and during overcast days. If the houses had been lived in it was probable that the effect of drawing heavy curtains at night could have made an appreciable difference to the result.

The investigation of solar space heating applications developed steadily from this point. The work at the Massachusetts Institute of Technology, which was started by the Godfrey L. Cabot bequest and which led to the building of Solar House No. I in 1940, continued with a series of different solar houses. Dr G. O. G. Löf, of the University of Colorado, was the earliest experimenter with solar air heaters, using a total collector area of approximately one third of the roof and passing the heated air either directly into the rooms or into a heat storage bin filled with crushed rock. The capacity of this store was about one full day's heating requirement and approximately 20% fuel savings resulted during the first season in 1946 (7, 8). Thirty years later by the beginning of 1976 the number of solar heated buildings which had been built since 1940 or which were under construction exceeded 200. Shurcliff's survey in March 1975 (9) included details of over 100 United States buildings but less than 20 for the rest of the world. By 1976 the UK total alone was approaching 20, with increasing encouragement from official government sources. By the end of 1977 Shurcliff confessed that the expansion in the number of solar houses had been so great that it was quite impossible to attempt to collect and cross-reference all the information. Three years later there were thousands of solar houses in the USA, and in Europe countries such as France had reported over 200.

A feature of the development of active systems during the 1970s and start of the 1980s was the increasing interest in applications with heat pumps.

The principles of the heat pump were established over a hundred years ago. By supplying energy to the heat pump, heat is transferred from a low temperature region to a higher temperature region. The earliest applications were in refrigeration, where the food is maintained at a temperature lower than the surroundings, while heat is rejected from the refrigerator to the surroundings by means of an external heat exchanger.

The coefficient of performance, COP, of a heat pump is defined as the ratio of the energy output to the input. The energy output appears as useful heat at a higher temperature than the surroundings while the energy input is supplied by electricity or the direct use of fossil fuels. The total energy input to the system includes the natural energy from the environment and most heat pump installations have a COP greater than 1.0. In other words, more useful energy at the higher temperature is obtained from the system than was supplied to it through the use of electricity or fossil fuels. Theoretically it is possible to obtain a COP in the order of 20, but in practice values of between 2 and 3 are normal for space heating applications in a temperate climate. The low temperature heat source could be soil, water or air. The use of solar energy to augment these low temperature sources of heat appears to be attractive because the higher the input temperature to the heat pump system becomes, the smaller the conventional energy input from electricity or fossil fuels for the same net energy output.

The various houses and buildings described in the following sections have been chosen to illustrate the historical development of solar heating leading to many current projects throughout the world. Starting with the USA, where much of the early work was carried out, applications are given from several other countries, including the UK.

SOME APPLICATIONS IN THE USA

MIT Solar House No. I (2, 9-11)

Constructed in 1939 as two rooms, an office and a laboratory, it had an approximate floor area of 46.5 m^2. The greater part of the roof sloped about 30° southward and contained 33.45 m^2 of exposed absorber surface placed in 37.9 m^2 of triple glazed collector. The absorber surface consisted of blackened copper sheet to which parallel copper tubes were soldered. The basement was filled with a large hot-water storage tank of 65.86 m^3 capacity with an average insulation thickness of 665 mm. In heat requirements, the building was designed to simulate a moderately insulated six-roomed house. It was the first 100% solar heated building, as it could store the summer's heat for winter use, but was regarded as very uneconomic and was demolished in 1941.

MIT Solar House No. II (9-12)

A single storey laboratory building constructed in 1947, its dimensions were approximately 4.26 × 13.4 × 2.44 m high with a bank of vertical solar collectors on the south wall consisting of seven different panels, each just under 10 m^2 in area. Various types of storage system were tested and in 1947-1949 it was converted into No. III.

MIT Solar House No. III (9-11)

With same floor dimensions as No. II, it had a roof-mounted, double-glazed collector with a similar absorber system to No. I, having an area of 37.2 m^2 at a tilt of 57° to the horizontal. Storage was in a 4.5 m^3 capacity cylindrical tank in the roof space. During the mid-winter four months up to 85% of the space heating was provided by the system and a mean annual value of about 90% was subsequently obtained. It was destroyed by fire in 1955.

MIT Solar House No. IV (9, 10, 13, 14)

Built in 1959, and illustrated in Fig. 4.1, it was considered unique (14) in
that it was designed as a solar house to make the fullest use of collected
energy, waste as little energy as possible and at the same time meet the
comfort and space requirements of modern living. It was a two-storey design
containing 134.7 m^2 of usable living area. The south elevation of the house
above ground consisted entirely of 59.5 m^2 of solar collector sloping at an
angle of 60° to the horizontal. The double-glazed collector was redesigned
and had the copper tube mechanically clipped onto blackened sheet aluminium,
the overall measured absorptivity being 0.97. The 5.7 m^3 water storage
tank was heavily insulated. In operation the occupants were careful not to
alter their individual ways of living to favour solar heating, so that dish
and clothes-washing operations were always carried out at the convenience of
the housewife and not only when the sun was shining. During the winter
season from 30 September 1959 to 30 March 1960, 44% of the space heating
load and 57% of the domestic hot water load was borne by the solar energy
system. This was rather less than the predicted performance, but was
considered to be due to the severe winter weather conditions experienced
that year. Maintenance problems caused the system to be abandoned after two
years, the overall performance for the two winters being 48% of the total
load supplied by the solar energy system.

The Dover House (3, 9, 10)

Claimed to be the first house heated entirely by solar energy, the solar
system was designed by Dr Maria Telkes, at that time a research associate
at MIT. A Boston architect, Eleanor Raymond, designed the house which is
illustrated in Fig. 4.2. It was built as a private project on the estate of
Amelia Peabody in Dover, Massachusetts and was first occupied on Christmas
eve in 1949. The double-glazed 66.89 m^2 air heating collector occupied the
entire vertical south face of the two storey building at second floor level.
Each collector panel consisted of two 3.28 × 1.22 m figured glass panes
separated by a 19 mm air gap. The absorber surface was made from standard
galvanized steel sheets painted with an ordinary matt black paint. Behind
each sheet was a 76 mm air space through which the air could circulate on its
way to three heat storage collector bins. These bins contained cans of
Glauber's salt - sodium sulphate decahydrate ($Na_2SO_4 10H_2O$) and occupied a
total volume of about 13.3 m^3. Dr Telkes felt that the difficulty of
storing heat for long periods using water or rocks lay in finding a large
enough space and suggested the use of the heat of fusion, or heat of melting
of chemical compounds. Glauber's salt, with a melting point of about 32°C,
could store about six or seven times more heat than water on an equal volume
basis. Only the ground floor area of 135.3 m^2 was heated and the heat was
transferred from storage into rooms by means of small fans which responded
to individual thermostats. The original storage capacity was estimated to
be about 12 full days mid-winter heating load. During the first year of
operation the solar system provided the entire heating load, but the
performance deteriorated after this, however, due to stratification and
irreversibility in the fusion and freezing of the salt, and auxiliary heat
was later required. The solar heating system was removed after four years
when the house was enlarged, but several very important design features had
by then been established:

Fig. 4.1.   MIT solar house No. IV.

Fig. 4.2.   The Dover house.

(i)   The effectiveness of the solar air collector, with its simple design
       and dual use as a heat collector and as a wall. This dual use of the
       collector as a wall or part of the roof was to feature in the great
       majority of subsequent solar house designs.
(ii)   The advantage of having controlled temperature zones in different
       parts of the house. This was overlooked in many later designs, but
       is once again being recognized as very important for saving energy.
(iii) The large heat storage capacity in a small volume which could be
       provided by heat-of-fusion salts. The problem of stratification with
       repeated cycling was still to be overcome and proved to be one of the
       most difficult problems in solar space heating applications until the
       developments outlined on pages 103-104.

The US Forest Service Bungalow (Bliss House) (15)

An existing single-storey bungalow at Amado, Arizona, with a floor area of
62.43 m^2 was modified in 1954-1955 to include a solar air heating system and
massive rock storage. The collector consisted of four layers of black
cotton cloth, with a 12.5 mm gap between each layer, under single glazing.
It had an area of 29.26 m^2 and was erected close to the bungalow at an angle
of 53° from the horizontal. The storage system was 65 tons, approximately
36.8 m^3, of 100 mm diameter rock, also located near the bungalow in an
insulated underground chamber. In operation, the air was circulated from
the collector into storage by means of a fan whenever the radiation
conditions were appropriate. A second fan supplied air to the house on
demand, either directly from the collector or from the storage system. The
system provided all the heat necessary for winter space heating in the
bungalow and was claimed to be the first 100% solar heated home in the USA.
For summer cooling, night air was drawn through a black cloth which covered
a separate horizontal bed. This cooled the air further - about 1°C - and it
was then passed to the storage system. During the day this cooled air could
be circulated to the bungalow. The system was demolished after just over one
year's successful operation.

An interesting design feature was that ten average days' storage capacity
was provided. This was more than adequate as one day's winter sun provided
over two days' heat demand. In more northerly latitudes much greater
storage capacity is necessary to compensate for the considerably lower
winter radiation levels. At that time the economics were adverse, as the
capital costs were over five times greater than a conventional heating system
and it was not possible to repay the capital costs and interest charges with
the amount of fuel saved - the ratio of capital cost to fuel saved being
50:1. It was also visually unattractive.

The Bridgers and Paxon Albuquerque Office Building (16, 17)

The world's first solar heated office building was built in Albuquerque, New
Mexico, and was first occupied in August 1956. The useful floor area was
approximately 400 m^2 and the building had south facing flat plate collectors
inclined at 60° to the horizontal as illustrated in Fig. 4.3. The net
collector area was about 70 m^2 with single glazing and there was a 22.7 m^3
underground insulated storage tank. All items of equipment in the building
were standard except for the collector plates, which consisted of aluminium
sheets, 0.476 mm thick, painted with a non-selective black paint and with
38 mm outside diameter copper pipes placed 150 mm apart, soldered to form a
continuous bond on the back side, and containing the heated water. Inside

Fig. 4.3. The Bridgers and Paxon Alubuquerque building.

the building, heat was supplied by passing the warm water from the storage
tank at about 40°C through pipes placed in floor and ceiling panels. A heat
pump was also used when storage water temperatures were not at a high enough
level to satisfy building heat requirements.

The advantages of using a heat pump with a solar collector were clearly
stated in the first report on the performance of the building (16). The
heat pump can be used for cooling in the summer season and this dual role is
attractive. In cold and cloudy weather the storage and collecting
temperatures may be allowed to fall very low with the resultant increase in
collector efficiencies and storage capacity. Use of the heat pump also
allows a smaller collector and storage tank to be installed. In the first
season of operation direct solar heating supplied 62.7% of the total heating
requirements while the remaining 37.3% was supplied with the heat pump
operating. It must be emphasized that the major source of heat was from the
solar collectors, even when the heat pump was operated. The amount of
energy required to run the heat pump supplied only 8.2% of the total heating
requirement. It is interesting to note the comment that with the cost of
fuels at that time (1956-1957) for heating, the savings in fuel costs did not
justify the necessary first cost expense for solar heating systems for most
localities in the USA. But there would be some areas where the high fuel
costs would make solar heating systems economically attractive. The system
operated in its original form for about six years with only occasional
trouble, for example when draining was incomplete and some frost damage
occurred. There was also deterioration in some rubber pipe hose connections.

The solar system was refurbished in 1974 as an ERDA project (17). The
principal changes are that the self-draining system has been replaced by an
ethylene glycol and water heat exchanger, pump and piping system to eliminate
the problems of freezing, and that five small water-to-air packaged heat
pumps have been added to extract energy from the circulating water in the
building and deliver heated air to five groups of rooms. The main objective
of the project was to develop generalised design data on solar energy
assisted heat pump systems for architects and consulting engineers.

The Thomason Houses (18-21)

The first house designed by Thomason was a single-storey house with a
basement and storage space underneath the pitched roof. Erected in
Washington DC in 1959 it had a collector area of 78 m² with a total living
area of 139 m². Thomason was one of the first designers to use the simple
and comparatively inexpensive 'trickle' collector system in which water is
pumped from a storage tank to a horizontal distribution pipe at the top of
the collector. In his original system, black corrugated aluminium was used
as the absorber surface and the collector had two cover plates, one of
glass and the other a transparent polyester film. The trickle effect was
obtained by allowing the water to flow through holes in the distribution
pipe directly opposite the channels in the corrugated sheet. An open
channel or gutter at the bottom of the absorber returned the heated water to
the storage tank. The storage tank consisted of a 6.1 m³ water tank
surrounded by 50 tons of small, 100 mm diameter, rock. The domestic water
system had a 1000 litre preheater. About five days' space heating storage
capacity was provided by the system and an overall performance of 95% solar
heating was claimed. For summer cooling, the water was directed over the
unglazed north facing roof channels during the night and was cooled by
evaporation, convection and radiation.

The second house, also erected in Washington DC in 1961, had a collector area of 52 m^2 and a heated living area of 63 m^2. Broadly similar in concept to the first house, it had an enhanced heating effect through a horizontal aluminium reflector surface of 31 m^2 extending from the base of the south facing collector. In his third Washington house erected in 1963, the storage tank was also used as a heated indoor swimming pool and the collector system was moved entirely onto the roof so that direct radiation from the winter sun could enter the living room and swimming pool windows on the south side A fourth house, with an inferior black asphalt shingle collector system was built next to the third, but was never fully tested and subsequently became a storeroom. Houses five to seven were described in 1973 (20), but only the sixth, a partially heated luxury house in Mexico City, was built.

The design for number seven included a shallow roof-pond collector with a reflector. Each night the heated water could drain to an underfloor heat storage area, where it could warm the floor and living space. In the mornings, a low-powered pump sent the water up to the roof. For summer cooling, the system could work in reverse, although the exact details were left for further design studies in any particular application. Two further houses were built in Prince George's County a few kilometres from Washington DC. In one of the houses (21), some modifications to the storage and collection systems tried in the earlier houses have been made. The main modification is that the rocks which surround the horizontal cylindrical 6.1 m^3 water storage tank can also be heated in winter by an oil-fired heater through a set of copper pipes. There are also two flue pipes from the boiler and in winter the exhaust gases pass through a flue pipe which also goes through the rock store.

The Thomason houses have been widely studied, as they provide a wealth of practical operating experience extending for nearly twenty years, and many of the early 1970s solar house designs were based on the Thomason systems. In the UK several trickle systems were built at that time but they have not been successful.

Solar One - The Institute of Energy Conservation, University of Delaware, USA (22, 23)

Solar One was built in 1973 as the first house to use a combination of thermal and photovoltaic solar energy in the same collector system. The other radical feature in the house is the extensive testing of heat of fusion thermal storage. Air is used to transfer heat from the collectors and a heat pump is available between 'hot' and 'cold' storage systems. The philosophy of the approach is summarised in a progress report (23) where it is pointed out that energy is required in different 'grades' in any one domestic housing application - low grade thermal energy for space heating or air conditioning, higher grade thermal energy for hot water, cooking and refrigeration, and light and electricity. Losses occur whenever one form of energy is converted to another and it is desirable to provide as much of these forms of energy as possible by converting them from the solar energy directly with any secondary conversion. As the necessary data for system optimization was not available in the early 1970s, the house was designed to allow maximum flexibility for experimentation. A cross-section through the house is shown in Fig. 4.4. The main single-storey living area contains the living/dining room, two bedrooms, bathrooms and kitchen area. The rear, north-facing, side contains a garage/exhibition area. As the house is designed to obtain performance data on each component in the system, to optimize the system and to improve the performance of the thermal-electric flat plate collectors, no attempt has been made to live in the house or to simulate occupancy. Details on overall performance, which would be of interest for an optimization of the collector: storage volume ratio and storage volume:house living area ratio, are necessarily limited by the extensive experimentation programme.

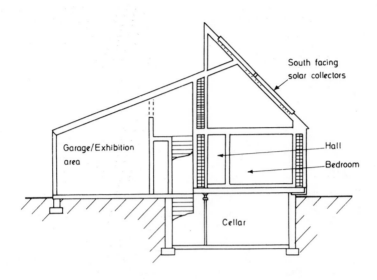

Fig. 4.4.   Solar One.

Twenty-four roof collectors, each nominally 1.2 × 2.43 m have been erected into the roof, which is inclined at 45° to the horizontal and faces 4.5° west of south. Three of these collectors are filled with cadmium sulphate/ copper sulphide (CdS/Cu_2S) solar cells manufactured between 1968 and 1970 by the Clevite-Gould Corporation. One hundred and four of these cells were connected in series in a sub-panel and three sub-panels fit into each main collector panel. Approximately 30 W of d.c. electrical power can be produced per square metre in full direct sunlight (about 3% efficiency). Air is circulated through the space underneath the solar cells and fins are used to improve the heat transfer characteristics to the air. Natural ventilation in the summer months is almost sufficient to keep the CdS/Cu_2S solar cells below their maximum operating temperature of $65^{\circ}C$. At temperatures of between $49^{\circ}C$ and $65^{\circ}C$, the thermal collection efficiency lies in the 50-70% range with ambient temperatures between $-18^{\circ}C$ and $10^{\circ}C$. A cross-section through a collector is shown in Fig. 4.5. By June 1975 a total of 16 different types of collector had been tested. All had similar glazing and collector casings, but various types of selective surface, fin spacing and geometry were investigated. As a further step in the simulation, a slaved power supply was used to produce the equivalent output from the whole thermal-electric roof, which has an effective area of 57.6 m^2. Six vertical south facing thermal air collectors, each nominally 1.2 × 1.83 m and originally planned with a simple selectively coated black aluminium sheet as the heat absorbing surface, were also included in the experimental programme.

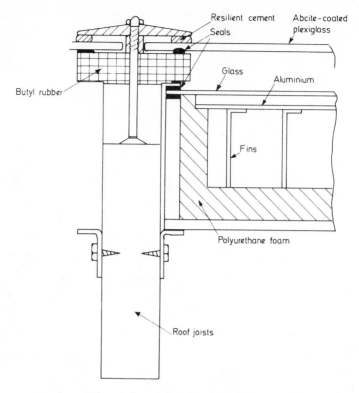

Fig. 4.5.   Solar One collector.

The heat storage system occupied a relatively small volume, approximately
6.12 m^3, and consisted of two outer vertical stacks of ABS pans containing
sodium thiosulphate pentahydrate ($Na_2S_2O_3.5H_2O$), with a change of state
temperature of 49°C.   The central stack contains a eutectic salt, mainly
sodium sulphate decahydrate ($Na_2SO_4.10H_2O$), with a change of state
temperature of 12.8°C, packed in tubes, each 31.75 mm in diameter and
1.83 m long.   The outer system is the 'hot' store while the central stack
is the 'cool' store.   Both these systems have been cycled extensively and
quite independently from the solar heating system.   The capacity of these
stores is sufficient to carry about three days' winter heat or one day's
cooling in summer.

The Copper Development Association Decade 80 House, Tucson (24)

This house was designed and constructed in 1975 as an industrial market
development project to create new applications for copper, brass and bronze.
The CDA claimed that this was the first 'real' home, as opposed to purely
experimental structures, to make so much use of solar radiation, as 100% of
the heating and up to 75% of the cooling load are predicted.   Although it
had many unusual features, such as the CDA electric vehicle in the garage
which is recharged nightly, the house was built to demonstrate that all the
essential components and materials needed to build an almost totally energy
self-sufficient home were already available at competitive prices.   A further

feature was that such a house could be built by any competent local building
contractor. After one year of full experimental trials the home was sold
for normal occupancy. Figure 4.6 shows the house with its integrated,
double-glazed, copper solar collector roof. The basic collector panel
consists of 1.2 × 2.44 m copper sheets laminated to plywood combined with
rectangular copper tubes for water circulation to the 11.4 m³insulated
storage tank. The CDA claimed that the solar roofing system would be paid
for in about ten years by savings in fuel.

Fig. 4.6. CDA decade 80 house, Tucson.

Cooling is provided by two standard lithium bromide-water absorption units
modified to use the solar-heated hot water as the heat source. Silicon
photovoltaic cells are also incorporated on the roof for various minor power
applications, such as the low voltage supply for a small TV set and a
kitchen clock. The solar cells also provide stand-by power for the home's
overall security system in case of electric failure.

The roof of the connecting guest wing, which is inclined at about 40° to the
horizontal, provides solar heating for the swimming pool in the spring and
autumn. In the summer it is used as a simple cooling system, as the pool
water can be circulated through it at night, reradiating to the cool night
sky, thus maintaining the pool at comfortable temperatures in daytime. The
main house roof is inclined at 27° to the horizontal to optimize on summer
heat collection to provide the relatively large amount of energy needed to

power the absorption cooling system. Further protection against unwanted
heat gain in the summer is provided by special solar bronze tinted double-
glazed windows on the side of the house facing the swimming pool.

The Henry Mathew House, Coos Bay, Oregon (25-27)

This house was designed and built by the owner, Henry Mathew, in 1966-1967
and is the best example of an owner-built solar house of its time. It
incorporates many basic design features which can be used as the starting
point in any solar housing system. It also has the classical simplicity of
the earliest solar houses with the living room and kitchen in the south side
to absorb the winter sun, but with a long overhanging roof to shade these
areas in the summer. Figure 4.7 shows the main features of the house and
its solar heating system. The roof collector, which is 15 m high × 24.4 m
long, is described in detail in Chapter 10, and has its performance improved
by the reflector, whch consists of standard household aluminium foil stuck
down with a conventional roofing compound. Water is pumped through the
pipes from the main storage tank by a 0.25 h.p. pump, controlled initially
by a roof-mounted thermostat. The pipes drain to a 170 litre surge tank
and then to the storage tank whenever the pump is not working. The storage
tank is insulated from the crawl space above it, but not from the earth
facing the sides and bottom. Substantial earth heat storage - and equally
heat loss - at certain times of the year is thus made possible and autumn
storage was found to occur in the 1974-1975 season. Insulated dampers in
the large heat ducts from the storage tank compartment to the living area
are thermostatically controlled and can cut off all heat in the summer.
There are no forced circulation air fans. Including the 30 m^3 steel storage
tank and approximately 37 m^2 of collector, the cost of the materials in 1967
was less than $1000. Constructing the tank, his first, took Henry Mathew
five weeks and the rest of the system another eight weeks. A further 30 m^2
of free-standing collector, erected some 20 m from the house, was added to
the system in January 1974. The atypical features of the house are as
follows:

(i)   It was built with quite standard components and not specially
       insulated although solar heating was envisaged from the start.
(ii)   Its location comparatively far north (42½°N) in a region noted for
       overcast cloudy winter conditions.
(iii) Its combination of nearly vertical solar collector (82° to the
       horizontal) and nearly horizontal reflecting surface (8° to the
       horizontal).
(iv)   The relatively large storage capacity of the 30 m^3 storage tank.
(v)   The combination of roof-mounted and free-standing collector, both with
       large reflecting surfaces immediately in front.

An unusual feature of the system is that no attempt is made during the summer
months to collect and store substantial amounts of energy for use in the
winter. Detailed results for the 1974-1975 season are available (25) and
show that about 85% of the total space heating demand was met by stored
solar energy. The Mathew family responded to the environment by allowing
their interior temperature to fall below the design value of 21°C so the
collector/storage contributions diminished as winter progressed.

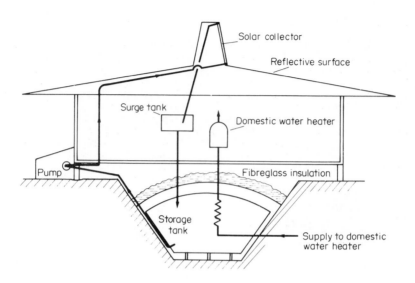

Fig. 4.7. Section through Henry Mathew house.

The Stationary Reflector/Tracking Absorber System (SRTA) (28)

The basic collector, which is described in greater detail in Chapter 6,
consists of a segment of a spherical mirror placed in a stationary position
facing the sun. It has a linear absorber which can track the image of the
sun by a simple pivoting motion about the centre of curvature of the
reflector. Figure 4.8 shows the dramatic impact the SRTA collector could
have on house design in regions where there are appreciable amounts of
winter sunshine. The radical feature of the system is that water can be
heated to a sufficiently high temperature for power generation and there is,
therefore, the possibility of self-sufficient systems without wind-energy or
direct electricity generation. Comparatively few details of any applications
have been recorded but a house incorporating a SRTA collector was designed
and built in Colorado in 1975.

With the high proportion of diffuse radiation in the UK, together with the
low levels of radiation in mid-winter, it is very unlikely that houses with
the SRTA system will be seen in the UK, although there would be every chance
of successful operation in the Mediterranean regions.

Fig. 4.8. The SRTA system.

SOME APPLICATIONS IN THE UK

The Cambridge Autarchic House Design (29, 30)

The project was initiated by the late Alex Pike of the Cambridge University Department of Architecture in 1971 with the objective of total autarchy, or self-sufficiency. Starting with the assumptions that the price of oil, gas, electricity and food may quadruple over the next ten years and that the demand for private space could increase to one acre per family and that the three-day working week could be the norm, he felt that these trends could stimulate families to want to produce their own power, water and food on site. Computer studies undertaken by Pike's Technical Research Division indicated that such a house was theoretically possible. A wind-driven generator was included in their model simulation as well as a solar collector occupying the entire south facing roof area, from which water could be returned to a 40 m^3 basement storage tank. A major design feature was the return to the concept of the Victorian Conservatory, with approximately half the total volume under the roof of the house consisting of a glazed patio area extending over the whole south facing elevation. During the very cold season this could be separated from the living areas and bedrooms by insulated shutters.

The work provided an extensive series of detailed preliminary system design simulations, but funds to take the scheme further were not forthcoming as Pike's original assumptions were not considered to be correct. The value of the work lay in the interest it generated among architects, engineers and all those who felt that something more should be done about the use of

alternative energy resources. Several of the ideas were widely adopted in
many other schemes, e.g. the Victorian conservatory, now called the solar
greenhouse, and the use of insulated shutters.

The First Milton Keynes Solar House (31-33)

In 1973 the Department of the Environment gave a grant for the design of an
experimental solar heating installation in the new town of Milton Keynes,
under the direction of Dr S. V. Szokolay, formerly of the Department of
Architecture at the Polytechnic of Central London. The aim of the project
was to test and prove the feasibility of solar heating in the UK. The
standard terrace house of the Milton Keynes Development Corporation, of
which hundreds are being built, is a thermally rather inefficient building
with reasonable insulation, but virtually no thermal inertia. It is felt
that future solar developments in the UK would use a more massive
construction and a far better thermal insulation, but with these advantages
a performance comparison with an identical house without a solar installation
could not be carried out. The solar house, illustrated in Fig. 4.9, which
was taken just prior to first occupation in March 1975, is fully
instrumented for continuous monitoring.

A feature of the design work was the extensive use made of the computer in
various parts of the process to simulate the hourly transmission of heat for
every day of the year. Although computer simulations were well known in
many applications in the USA, this was the earliest practical example in the
UK. It was predicted that in the period from April to September the entire
demand for space heating would be satisfied, and only in December and
January would the percentage of demand satisfied fall below 30%. For
domestic water heating, values from April to September were between 70% and
85% of the demand satisfied, although naturally there was a considerable
fall off in the winter months. The very long hot summer of 1975 caused
overheating problems in the bedrooms, and it was first thought that this was
caused by the temperature in the storage tanks immediately adjacent to
the bedrooms reaching 70°C. However, similar overheating problems occurred
in other adjacent non-solar heated houses, so this is probably a basic
design feature of the whole terrace. Full details of the final design give
details of the 30° sloping roof (computer simulation called for 34°) with
the 36 m^2 of solar collector (design requirements based on 40 m^2). The
total floor area is 90 m^2. The capacity of the storage tank is effectively
4.2 m^3 (design called for 5.2 m^3) and it is insulated with 100 mm of glass
fibre insulation.

By 1981, very considerable experience had been obtained from the continuous
programme of monitoring under the direction of Professor Ray Maw of the
Polytechnic of Central London. The house has been normally occupied and
dissipates 180 W/K of total space and water heating or 14 MWh annually.
The thermal energy balance for 1979-1980 (33) is shown in Table 4.1.

Excluding passive and occupancy gains the active solar system supplied 58%
of the total heating plant load for the house in 1979-1980. The original
system, which produced only 37% of the required load, had been extensively
modified over the years to achieve this performance. The measures taken
included maximizing the use of solar heated water, down to 23°C; increasing
the room air-heater size; adding anti-freeze to obviate warming of the
panels in freezing conditions and limiting the immersion heat auxiliary
supply temperature to 50°C.

Fig. 4.9. The Milton Keynes house.

Table 4.1.   Thermal energy balance in the PCL solar house
             (1979-1980)

Primary supply	kWh	Useful consumption	
Passive solar gain (computed)		3858⎱	50% of total consumption
Occupancy gain (computed)		3060⎰	
Active solar space heat	2036	2036⎫	
non-useful losses	4496		59% of total plant load
useful losses	388	388⎬	
Active solar water heating	1822	1822⎭	
Gas space heat auxiliary	1993	1993	
pilot and jet waste	2291		
Electric water auxiliary	730	730	
cylinder losses	195		
fan and pump	491		
(generation + supply loss	7420)		
Total plant output	14442	6969	

The cost of the system in 1974, net of tiling and radiators displaced, was
£2290, or £4900 in 1981 prices. It saved £110 per year in gas space and
electric water heating bills in 1981. By assuming a doubling of gas and
electricity prices over the life of the system (say 20 years) and that the
system was paid for with a tax-deductable mortgage as part of the house,
Maw and his colleagues concluded that it would have paid for itself in
18 years, allowing for scrap value of £300.

While this conclusion may not sound particularly impressive, the importance
of this project should not be underestimated. By the early 1980s Milton
Keynes has been established as the leading area for research work in many
types of solar building in the UK, with hundreds of houses, active and
passive, being monitored by several different research groups.

The Granada House, Macclesfield

In January 1976 the Granada Television Company, Manchester, presented a
series of programmes on the conversion of an old coach-house into a four-
bedroomed solar heated house. Many other energy saving ideas which feature
in the most sophisticated solar housing research investigations were also
considered, such as the recovery of waste heat from the hot water and a
ventilation heat recovery system. Various types of insulation have been
tried, including 50 mm of glass fibre blanket with 50 mm expanded
polystyrene slab, 100 mm of ordinary glass fibre or 100 mm semi-rigid
mineral wool slabs on different sides faced with timber planks and weatherboarding. Detailed accounts of the construction (34) and the subsequent
initial 'settling down' period (35) have been published. Based on 1975 UK
Building Regulations, the gross annual space heating requirements for the
house would be 45 230 kWh, but careful attention to double-glazing and
ventilation as well as insulation reduced this to 21 910 kWh. It was
originally thought that this demand could be met as shown in Fig. 4.10,
which is the theoretical annual house energy diagram based on average
weather conditions, prepared by the Electricity Council Research Centre,
Capenhurst. The shaded area represents the net supplementary space
heating required, 3680 kWh, with an internal living space temperature of

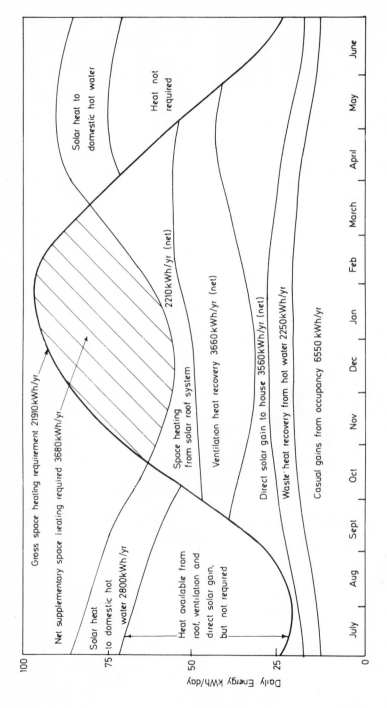

Fig. 4.10. House energy diagram through the year based on average weather conditions.

19.5°C, and an overall solar collector efficiency of 30%. The contribution
from the other heat sources to the gross space heating requirement were
calculated in 1975 and are shown in Table 4.2.

Table 4.2.

Source	kWh/annum
Space heating from solar roof system	2210
Ventilation heat recovery	3660
Direct solar gain to house	3560
Waste heat recovery from hot water	2250
Casual gains from occupancy (cooking, lighting, etc.)	6550
Total	18230

The U-values for the house are compared with the original U-values and those
for housing at 1978 standards in Table 4.3.

Table 4.3.

U-values, W/m^2 K

Building component	Typical range since 1945	Maximum under 1978 building regulations	Low energy house	
			Original	Converted
Walls	1.0-1.7	1.0⎱ 1.8 average	2.0	0.36
Windows	4.3-5.7	5.7⎰	4.3	2.5
Roof	0.85-1.7	0.6	1.7	0.22
Floor	0.6-1.2	1.0	0.75	0.36

Figure 4.11 shows work on the solar roof (south-west) and the north-western
side of the house with the slate-tiled outhouse, which also contains the
3000 litre solar heated water storage tank and a 2000 litre dump tank.
This illustration was taken as the solar roof was being laid. The roof,
which has a collecting area of about 42 m^2, is based on the Thomason
concept and made of standard corrugated aluminium painted with an acrylic
matt black paint, single glazed with 4 mm horticultural glass. Water
trickles down the channels from a horizontal perforated pipe laid just
below the ridge of the roof.

The north-western side of the house has only one window, while the long
north-eastern side has only three. A feature of the house is the large
conservatory at the ground-floor level on the south-western side. Heated
air from this conservatory can pass directly into the top floor of the house
through ducting at the bottom of the first-floor bedroom windows.

During the year from September 1977 to August 1978 the building was
monitored by the Electricity Council Research Centre (36) and the total
energy utilization, shown in Table 4.4, presents a rather different picture
from the projection shown in Fig. 4.10 although the predicted total load was
almost 100% correct.

Table 4.4

	kWh
Solar energy from roof	318
Direct solar gain	4642
Solid fuel	6231
Domestic electricity including heat pump and immersion heater	9155
Casual gains from occupants	1412
Total	21 758

Fig. 4.11.   The Granada house.

The conclusions drawn by the Electricity Council after their year's work were as follows:

(1) That low energy requirements result in low solar energy contribution.
(2) That a heat pump greatly improved the solar panel collection efficiency.
(3) That the fuel system used gives a cost per kWh comparable with off-peak electricity.
(4) That large thermal mass and good insulation improve comfort levels by reducing temperature swings.
(5) That the low temperature water return system gave very satisfactory results.
(6) The mechanical and natural ventilation systems were very unobtrusive and liked by the occupants.

The sponsors of the project, the Granada Television Company, and the system advisors at the Electricity Council Research Centre were clearly influenced by the simple Thomason trickle roof concept. The decision to use this type of roof turned out to be wrong, as a glance at Fig. 3.24 could have easily predicted. With anything other than excellent radiation conditions or a requirement for a small temperature rise, the trickle system has a relatively poor performance. But, apart from this one major drawback, the house has been an outstanding success, particularly from the viewpoint of the occupants, who enjoy living in an extremely attractive house with the added advantage of very low heating costs.

The Building Research Establishment Houses (37)

Three experimental houses have been built at the Building Research Station, Watford, to study three major techniques of energy conservation – solar energy, the heat pump and waste heat reclamation. In marked contrast to the Philips concept, which is described later, the three different sets of options can be studied simultaneously and the BRE feel that there is no single 'best' universal solution. The performance of the houses are monitored under controlled conditions with simulated occupancy. The solar house (and the waste heat reclamation house) are based on a timber-framed, five person, two-storey house, the Bretton Type 47, already studied extensively by the BRE in a district heating scheme at Bretton, Peterborough. The timber structure of these houses is prefabricated and the exterior cladding is of brick and weatherboarding. A thickness of 92 mm of insulation in the roof and external wall panels gives a U-value of approximately 0.29 W/m^2. The roof of the solar house is pitched at 42° to the horizontal to give a better all-round performance compared with the 22$\frac{1}{2}^\circ$ of the standard Bretton Type 47. The energy system of the solar house is shown in Fig. 4.12. The preliminary design details included a 22 m^2 solar roof and a 35 m^3 well-insulated tank located outside and below ground level. Space heating is by radiators, but these are sized larger than normal so that lower water temperatures can be used. Various operating modes are selected according to the prevailing conditions. When the 35 m^3 storage tank is at a sufficiently high temperature, the radiators can draw their heat from it. At other times they draw their heat from an insulated 1 m^3 tank which, in turn, is heated by a small off-peak electric heat pump using the 35 m^3 storage tank as its low temperature source. The domestic hot water system is fed from a 300 litre storage tank, sufficient for 24 hours of normal use. This tank is heated either from a heat exchanger in the 35 m^3 tank or through another small off-peak heat pump. The unusual feature about the solar collector system is that energy can be transferred to the

35 m^3 storage tank even when the temperature from the collector is lower
than the tank temperature. This is achieved by the use of another heat
pump.

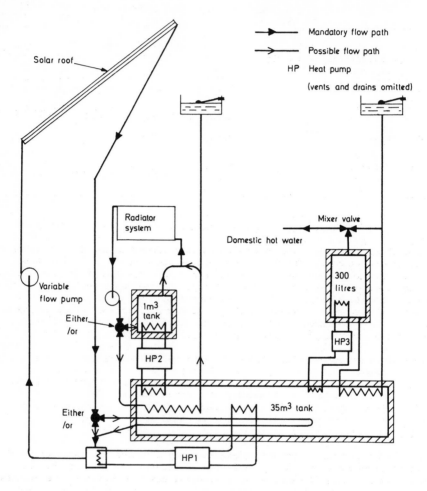

Fig. 4.12.   BRE house energy system.

The comparative annual energy consumptions predicted for the three houses
are shown in Table 4.5 together with figures for a standard Bretton Type 47
and a specially insulated house.

The derivation of primary energy is obtained by making conventional
assumptions on overheads and utilisation efficiencies for electricity, gas
and oil. The relatively high values of primary energy in both the heat pump
house and the solar house are due to the almost exclusive use of electricity
to provide the net energy.

Table 4.5. Energy consumption in GJ per annum

	Heating	Net energy	Total net energy	Primary energy
Bretton Type 47 (current building regulations)	Space Water	54.0 12.0	66.0	151.8
Bretton Type 47 (0.29 U-value)	Space Water	27.0 12.0	39.0	89.7
Heat reclaim house	Space Water	21.0 6.5	27.5	54.4
Heat pump house	Space Water	9.0 5.0	14.0	50.1
Solar house	Space Water	13.5	13.5	50.0

The Philips Minimum Energy House, Germany (38, 39)

An analysis of the energy consumption in the Federal German Republic showed
that about half was in the form of low temperature heat, defined as heat
available at less than 100°C. The major part of this low grade heat was used
in the private sector for heating buildings and providing hot water - a
pattern broadly similar to that in many northern European industrial
countries. The Philips research programme concentrated on this area and has
identified the measures which can be taken to reduce the consumption of
conventional sources of energy into four groups as follows:

(i)   the reduction of heat losses through floors, ceilings, walls and
       windows;
(ii)   the recovery of waste heat from the various domestic water systems
       and the exhaust air from the ventilation system;
(iii) the use of alternative energy sources which are not harmful to the
       environment, e.g. heat from the soil and solar energy; and
(iv)   the development of optimized integrated energy systems.

The aim of the research programme was to study the economic feasibility of
these resources and to develop optimized integrated energy systems. The
experimental house, illustrated in Fig. 4.13, has been built in the grounds
of the Philips Research Laboratory in Aachen. All the design parameters
such as its size, furnishing and household appliances were selected to match
the requirements of an average German family of four. Two Philips P855
process computers simulated the energy demands of the family as well as
controlling the various systems and processing all the data. The various
major design features are shown in Fig. 4.14. The emphasis is on flexibility
and many different combinations of solar heating and storage at different
temperature regimes are possible, as well as the combinations of the heat
pump system with the waste water and/or the heat from the earth. Table 4.6
gives some basic data about the house.

Table 4.6.

Some basic data	
Cellar area	150 m^2
Living area	116 m^2
Window area	23.5 m^2
Living area volume	290 m^3

Long term heat storage unit	
Volume	42 m^3
Insulation	250 mm rock wool
Temperature range	5–95°C

Domestic hot water storage unit	
Volume	4 m^3
Insulation	250 mm rock wool
Temperature range	45–55°C

Waste water tank	
Volume	1 m^3
Insulation	100 mm rock wool

The connected load of the electrical heat pump is 1.2 kW and at a temperature range between 15°C and 50°C its rated coefficient of performance lies between 3.5 and 4.0.

The effect of providing extra insulation for the walls, floors and ceilings, reducing the ventilation losses and providing specially coated double-glazed windows is shown in Table 4.7. Compared with a normal house, the overall thermal losses are reduced by a factor of six, and compared with a well insulated house by a factor of three.

Table 4.7.

	Average house		Well-insulated house		Experimental house	
	U_L (W/m^2 K)	kWh/year	U_L (W/m^2 K)	kWh/year	U_L (W/m^2 K)	kWh/year
Walls, floors, ceilings	1.23	32 630	0.48	12 600	0.14	3 630
Windows	5.80	9 970	3.3	5 700	1.5	2 570
Leakage	–	7 000	–	7 000	–	700
Controlled ventilation	–	–	–	–	–	1 400
Totals		49 600		25 300		8 300

Fig. 4.13. The Philips house.

For leakage rates it was assumed that there was one air change per house;
for controlled ventilation, one air change per house with 80% heat recovery.
The average annual energy use assumptions for a family of four were given as
follows:

(i)   hot water, washing machine and dryer, dish washer - 3980 kWh;
(ii) deep freeze, refrigerator - 1095 kWh;
(iii) lighting, television and other appliances - 1820 kWh.

This gives a total of 6895 kWh, but with the waste heat recovery heat pump
system, only a small percentage of the 3980 kWh used for hot water needs to
be supplied as external electricity. A coefficient of performance of about 3
is sufficient to save 3000 kWh, making a net demand 3895 kWh.

The energy from the earth can be used for both heating and cooling. In the
heating mode, a heat exchanger consisting of a 120 m water filled plastic
pipe was installed under the 150 m^2 cellar floor and by using the 1.2 kW
heat pump, heat can be transferred from the soil, which has a temperature of
about $7^{\circ}C$, to the hot water tank at a temperature of $50^{\circ}C$. Cooling is
supplied with hardly any extra energy expenditure, as air is drawn in through
a hollow cinder brick wall at cellar level. The solar collectors are
integrated into the south facing roof, as shown in Fig. 4.13 and are inclined
at an angle of 48° to the horizontal and cover an area of 20 m^2. Each of
the 18 collector boxes contains 18 tubular evacuated glass tubes which were
described in the previous chapter. The computer predictions indicated that
the 10 m^2 of collector would collect between 10 000 and 12 000 kWh annually,
which is more than the total heating needs of the house.

92   Sun Power

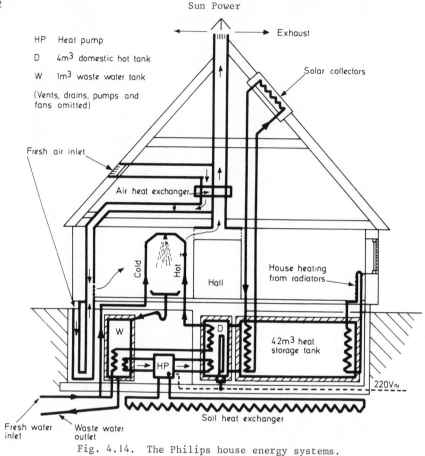

HP Heat pump

D   4m³ domestic hot tank

W   1m³ waste water tank

(Vents, drains, pumps and
fans omitted)

Fresh air inlet

Air heat exchanger

Cold   Hot

Hall

House heating
from radiators

W

D

42m³ heat
storage tank

HP

220V∿

Soil heat exchanger

Fresh water   Waste water
inlet   outlet

Exhaust

Solar collectors

Fig. 4.14. The Philips house energy systems.

After some five years of continuous monitoring and assessment of the results,
a number of broad conclusions emerged. The first was that investment in
insulation is more sensible than attempting to put an active solar space
heating system into a relatively uninsulated building. Waste heat recovery
from both the waste water and the ventilation system were found to be both
practical and economic. Should all the Philips systems be adopted in new
housing, the annual energy demand could be reduced by a factor of ten, with
a maximum space heating requirement of some 2 kW (in a country such as
Germany). They also concluded that large water or rock thermal storage
systems did not appear to be economic for single houses but that multiple
housing schemes with a control collector system linked to a common store
could be a possible solution - an approach already being followed in several
countries. The hybrid approach, with passive space heating and active
domestic water heating was also favoured.

Heat Pump and Optical Systems

A limited proportion of the space heating demand can be obtained simply by installing a fan system, controlled by a differential temperature controller, in an existing, unglazed attic space. More sophisticated roof systems make use of the characteristics of conventional roof slopes by replacing a considerable area of south-facing roof by glazing, thus allowing radiation to penetrate into the attic space. Both types of system were first studied during the 1970s. One transfers the heat in the attic directly to the rest of the house by means of a heat pump, the other uses an inexpensive reflecting optical system.

Heat Pump System (40)

In the University of Nebraska (Lincoln) and Lincoln Electric System solar house a south-facing glazed roof allows heat to be reflected in the attic, where natural circulation drives the heated air to the apex of the roof. A conventional outdoor heat pump is located near the apex and has controls which allow heat to be extracted either from the outside or from the attic. This heat is circulated to the water storage tank, which is maintained at a minimum temperature of $40^{\circ}C$ using auxiliary heat where necessary. A vertical duct also allows direct circulation of attic air to heat the house. While the heat pump is collecting energy from the attic, the temperature there is maintained at approximately $10^{\circ}C$, thus high solar collection efficiencies are achieved without expensive double glazing. The stored hot water is pumped through a fan coil unit serving a conventional ducted warm air heating system. For summer cooling the heat pump is reversed, storing cold water at a minimum of $5^{\circ}C$. The use of conventional, commercially available systems and materials forms part of the project, which is intended to demonstrate the economic viability of the design for typical American mid-west climatic conditions, where approximately 60% of the winter radiation is direct. One of the economic aspects thrown up by the analysis is that although there is a relatively higher electrical consumption during solar collection, this is more than offset by the greatly reduced capital cost of the system when compared with existing flat plate collector systems. The projected design studies indicated that an overall coefficient of performance for the heat pump would be 2.72 based on a total of 800 hours' operation compared with a value of 1.70 obtained from a typical conventional system installed in Lincoln.

Reflecting Optical System (41)

One system which has been described uses only plane reflecting surfaces and focuses with direct and diffuse radiation from a relatively large entrance area onto a flat plate collector typically about one fifth of the entrance area. The principles of the pyramidal optics system are shown in Fig. 4.15.

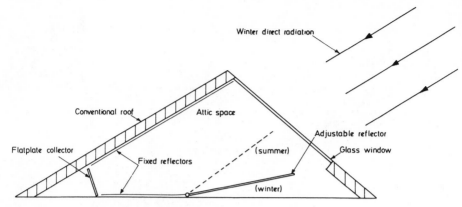

Fig. 4.15.   Pyramidal optical system.

The system consists of stationary flat reflecting surfaces which make up a
focusing two-dimensional pyramid and a moving flat reflective surface which
is adjusted for periodic variations in the sun's path - usually seasonal.
A net optical gain ranging from 1.6 to 4.8 has been claimed for the system
and its potential for achieving high temperatures from conventional flat
plate collectors could be of considerable importance in absorption cooling
applications. Various combinations of moving reflector are possible,
including an externally mounted system which was used in a prototype
installation at Stanford, Connecticut.

A major advantage of all glazed solar roof systems is the attractive
appearance which the houses can have using conventional design and
construction techniques as the attic solar collector can be taken for a
normal bedroom window, as illustrated in Fig. 4.16.

Other Heat Pump Systems

An interesting application I studied was a solar heated water system linked
to a heat pump for the visitor centre at Clumber Park, Nottinghamshire (42),
which showed that the old underground water tanks in the grenhouses could
have a new life as the heat storage system.

Studies on the combination of heat pumps and roof collectors which started
in Germany during the 1970s resulted in the Rheinisch-Westfälisches
Elektrizitätswerk (RWE) encouraging the use of a series of different low-
temperature collector systems (43). To illustrate their approach, they
pointed out that a surface of only 1 m^2 with a vertical air stream passing
through it with a very low wind speed of 1 m/s would have a heat extraction
capacity of 1.3 kW if the air stream were cooled by only 1°C. This concept
was extended to a range of unglazed solar collector systems, of which the
best known was the "Energie-Dach", or energy roof. Nevertheless the systems
were not entirely accepted as being superior to conventional heat-pump
systems, especially when the extra costs of the energy roof were taken into
account.

Fig. 4.16. Attic space heating.

ANALYSIS OF SOLAR SPACE HEATING SYSTEMS

In a review (44) of Shurcliff's "Solar Heated Buildings - A Brief Survey"
(9), the comment was made that the variety of designs described will
fascinate the reader and thoroughly confuse the earnest seeker after the
optimum system. One of the difficulties in attempting any analysis is that
even with conventionally heated houses, apparently similar families in the
same district, in almost identical houses, will have wide variations in their
heating costs. An assumption, which is probably very reasonable, has to be
made that all solar houses included in the analysis are occupied by fairly
similar families who will try to get as much use of solar energy as
possible out of their particular system. A second difficulty is trying to
define what is meant by 'per cent solar heated'. Where evidence on how a
family behaves in a solar house is available, such as in the Henry Mathew
house (25), it appears that there is an acceptance of the prevailing
interior temperature, even when this falls below a level which conventional
heating might have been expected to maintain. This makes a precise assessment of how much conventional heating is actually needed rather difficult.

The main factors which could be considered in the analysis are as follows:

   (i)   ratio of collector area to floor area;
   (ii)   position, angle of inclination and type of solar collector;
   (iii) ratio of storage volume to floor area;
   (iv)   type of storage system;
   (v)   geographical location of building;
   (vi)   overall insulation features, including window size, orientation and
          glazing; and
   (vii) height of the heated rooms.

Some further assumptions can now be made. The storage system can be based
on an equivalent volume of water. The overall insulation and height of the
heated rooms is hardly ever included in sufficient detail for analysis, so
variations must be neglected. This means that at any particular latitude a
series of curves could be drawn, giving the 'per cent solar heated' value
with various ratios of collector area/floor area plotted against the
storage volume/floor area ratio. This has been used as the basis of the
analysis. Fig. 4.17, showing a few curves based on latitudes less than
40°N, illustrates the main features.

Early solar houses were generally not very well insulated and their solar
collector systems were less efficient than current designs, so the lowest
curve on Fig. 4.17, with a collector area/floor area ratio of 0.6, represents
the performances which were achieved in the 1950s. Improvements in thermal
insulation and in collector systems since then have made a considerable
difference to the performance predictions. The main tendency is that
greater 'per cent solar heated' values can be obtained with relatively
smaller collectors and storage systems as shown by the two upper curves.
Considering one particular case, a performance originally represented by
point A. A slightly smaller collector area/floor area ratio would give a
100% solar heated performance now at point B. If the same level of overall
performance were required, both the collector area and storage volume could
be halved to give point C.

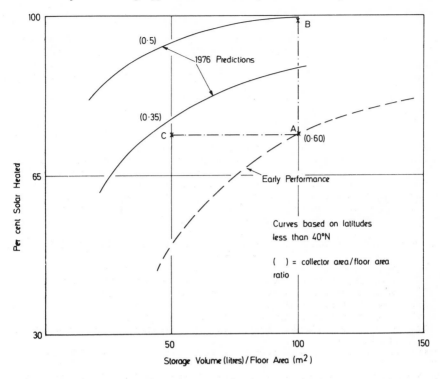

Fig. 4.17.

Figure 4.18 shows actual and predicted performance points for various solar buildings at latitudes less than 40°N. With storage volume (litres)/floor area (m²) ratios greater than 100 it can be seen that 100% solar heating is quite possible and that very high values approaching 90% are predicted for quite low ratios of both storage volume and of collector area to floor area.

The comparison between Fig. 4.18 and Fig. 4.19, which gives the points for latitudes greater than 40°N, is interesting as immediately the amount of solar heating which can be provided is seen to be smaller. Only the Henry Mathew house, with a significantly large volume of storage, stands out with a relatively small collector area/floor area ratio of 0.44. In both Fig. 4.18 and Fig. 4.19 the predicted performances, with the improved insulation standards and collector performance, are all following the trends illustrated in Fig. 4.17.

By the end of the 1970s considerable experience had been obtained from the relatively small number of houses which had been designed to use active solar systems to provide 100% space heating in relatively cold climates, typically greater than 3000°C-days. Four of these houses were studied in detail by Besant and Dumont (45) and their main features are shown in Table 4.8.

Table 4.8. Active solar system 100% space heated houses.

	MIT Solar I Cambridge, Mass. 1939-1941	Zero-Energy House Denmark 1975-	Provident House Toronto 1976-	Saskatchewan Conservation House Regina 1977-
Heated floor area (m²)	46.5	120	325	188
Solar collector area (m²)	33.5	42.0	66.6	17.9
Collector type	Water, flat plate 3-glazings non-selective	Water, flat plate 2-glazings non-selective	Water, flat plate Single-glazing selective surface	Anti-freeze evacuated tube selective surface
Collector tilt angle	Latitude -12°	Latitude +34°	Latitude +12°	Latitude +20°
Collector area/floor area	0.72	0.35	0.20	0.10
Water storage volume (litres)	79 000 steel tank	30 000 steel tank	272 000 concrete tank	12 700 steel tank
Water storage volume/house volume	0.71	0.10	0.35	0.028
Annual solar radiation on horizontal surface (GJ/m²)	5.0	3.5	5.0	5.1
Latitude	42° 20'	55° 41'	43° 40'	50° 30'
Annual degree days (°C-days) (Reference 18.3°C)	3130	3738	3793	6003
Thermal resistance values (m² °C/W)				
Ceiling	2.6 (R15)	9.2 (R52)	7.0 (R40)	10.6 (R60)
Walls	2.1 (R12)	6.9 (R39)	4.9 (R28)	7.3 (R41)
House heat loss coefficient (W/°C) Heat loss rate per unit temp. diff. between outside and inside	108	119 shutters open 75 shutters closed	272	99 shutters open 68 shutters closed
Insulating window shutters	No	Yes	No	Yes
Air-to-air heat exchanger	No	Yes	Yes	Yes
Waste water heat exchanger	No	Yes	No	Yes

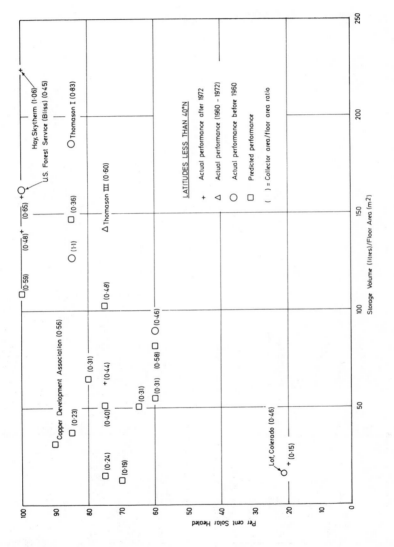

Fig. 4.18.

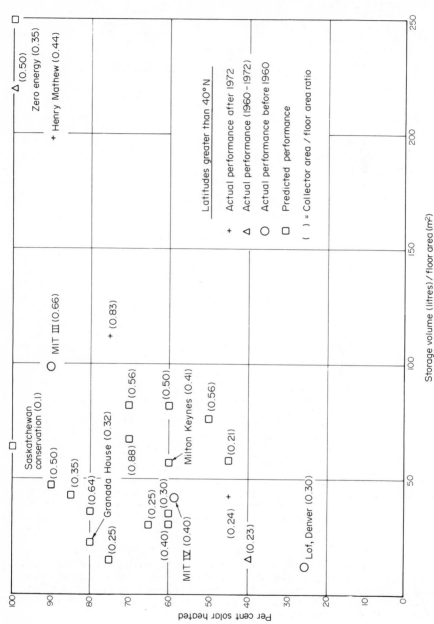

Fig. 4.19.

Only two of the four houses, the Zero Energy House and the Saskatchewan Conservation house, can be included in Fig. 4.19 as the storage volume/floor area ratio of the others was too large. The interesting feature of this analysis is that the trends in system design to achieve a high percentage of solar space heating which were beginning to be appreciated during the early 1970s, and which can be seen in the buildings shown in Figs 4.18 and 4.19, can now be confirmed.

Besant and Dumont (45) summarized their findings as follows: "To achieve 100% solar space heating in a relatively harsh climate (>3000°C-days) with relatively small collector areas and small storage volumes, it is likely that the following features should be incorporated in such dwellings.

(1) Use of high standards of energy conservation
     (a) well-installed vapour barrier and an air-to-air heat exchanger;
     (b) insulation levels that substantially exceed current building standards;
     (c) thermal shutter systems for windows.
(2) Use of passive gain through orientation of windows to the equator.
(3) Use of high efficiency solar collectors."

Current building standards insulation levels in many countries are often considered to be inadequate by those who have studied the potential benefits of low energy housing.

EXPERIENCE WITH PERFORMANCE SIMULATION, MONITORING AND CONTROL STRATEGIES

The first priority of the Performance Monitoring Group of the Commission of the European Communities was to develop a standard format for reporting results. They had found that early data were extremely difficult to analyse, as information from the different projects was presented in a variety of ways with widely differing terms and definitions. By 1981 30 formats had been completed and details of collaborative work in field trials in the eight participating countries has been published (46).

Computer simulation techniques have developed in parallel with the growth of solar systems in the past decade. TRNSYS (47) is probably the best known modular solar process simulation program in the world, with subroutines that represent all the significant components in typical solar energy systems. Current listings are available from the University of Wisconsin-Madison. Work leading to the present comprehensive list has been widely published, typical examples being the rock bed storage and discharge flow simulation which was compared with a real system for a period of two weeks' normal system operation (48) and the seasonal storage of energy (49).

Long-term collector performance has been modelled with considerable accuracy using graphical data and equivalent algebraic equations to determine monthly average collection efficiency (50). This is claimed to provide a quick method for comparing different collector designs in different locations with a variety of inlet temperatures.

A detailed examination of control strategies would be beyond the scope of this book. However, a good example of this work, reported by Jesch *et al.* at the University of Birmingham (51) was the re-routing of the stored water in separate tanks so that the external heat exchanges from the closed primary circuit were always working with the largest available temperature difference.

THERMAL ENERGY STORAGE

Thermal energy storage is essential for both domestic water and spaceheating applications and for the high temperature storage systems needed for thermal power applications. There are other applications, such as horticultural or in the process industries, where a storage facility is also required. For certain applications, particularly the need for space cooling during summer months, it is an advantage if the storage unit can also store at low temperatures. The choice of storage material depends on the particular application and for many domestic applications water and/or rock storage systems have been developed. A combined solar air heater/rock storage structure is illustrated in Fig. 4.20. First described in 1974 (52), it is a fully portable A-framed insulated unit containing washed river gravel with the air heater on the south facing sloping wall assisted by a hinged reflecting surface, which can be used to cover the collector at night to prevent heat losses. Studies on the use of gravel bed storage have been carried out for several years in Australia (53-55) and the advantages of replacing gravel with a strongly adsorbing material such as silica gel or activated alumina are discussed by Close and Pryor (56).

Water and rocks are typical examples of materials which store energy as specific heat (sensible heat), but their use is limited by their comparatively low specific heats. The heat of fusion (latent heat) which is involved when a substance changes state from a solid to a liquid provides an attractive method of storing a given amount of heat within a much smaller volume. This is illustrated in Table 4.9, which shows the weight and volume of rock-like solids, water and heat of fusion materials for storing 1 GJ (about 278 kWh) with a 20°C temperature limit. The table is based on data originally given by Telkes (57), who discussed the properties of a wide range of salt hydrates which could be used for heat storage. The least expensive and most readily available is sodium sulphate decahydrate ($Na_2SO_4.10H_2O$) or Glauber's salt, which has to be mixed with 3-4% borax as a nucleating agent if complete crystallization is to be obtained. These processes occur at about 30°C.

Table 4.9. Thermal storage of 1 GJ

	Rock	Water	Heat of fusion materials
Specific heat, kJ/kg $^{\circ}$C	0.837	4.187	2.09
Heat of fusion, kJ/kg	–	–	232.6
Density, kg/m^3	2242	1000	1602
Storage of 1 GJ, weight, kg	59737	11941	3644
Relative weight	16.4	3.27	1
Volume, m^3	26.6	11.941	2.274
Relative volume	11.69	5.25	1

Fig. 4.20.

However, practical commercial systems were not fully developed until the early 1980s.

The problem was that the crystallization process always started at the interface between the saturated liquid and previously precipitated solids and a barrier was quickly formed between liquid and solid phases. With each cycle, less 'stored' energy could be recovered. Many different chemical approaches to the problem have been tried and this work has been clearly described by J. K. R. Page and his colleagues of the Calor group (58). Their commercial process was still subject to patent applications but they indicated that in addition to the 4% borax nucleator they had added approximately 8% of polymeric stabiliser. To store 135 kWh their storage unit occupied about a quarter of the volume of a practical water storage system.

A particularly useful polymer system for these applications was identified as

$$([CH.CONH_2]_x \cdot [CH.R^-]_y)_n.$$

The use of unthickened sodium sulphate decahydrate in layers less than 65 mm thick was first shown by Telkes (57) to be a possible means of overcoming the problems of crystallizing during multiple cycling. This principle was used by Johnson (59) at MIT to develop a ceiling tile in which these thin layers could be placed. The tiles were made out of a polyester concrete (approximately 15% polyester resin and 85% aggregate). A building on the campus was successfully heated by the reflected radiation from the white walls. By 1981 these tubes were commercially available in the USA.

Normal paraffins have been studied extensively as possible thermal energy storage systems. A series of compounds which have broadly similar properties but with the advantage of remaining solid after the phase change, known as 'layer perovskites' have been identified in Italy (60) and could prove to be a useful additional storage medium during the 1980s.

For high temperature storage, in the order of 200-300°C, other salts have been considered (61, 62) and the heat of hydration of inorganic oxides, principally MgO and CaO (63).

Interseasonal Storage

The different options for the long term interseasonal storage of solar energy have been analysed by Givoni (64). He considered the storage media, together with combinations of heat transfer media, and produced a table which gives the optimal combinations. These are reproduced in Table 4.10, where a plus sign marks a realizable combination and a minus sign an improbable one.

Table 4.10.

Storage media \ Heat transport media	Water	Air	Water + air	Steam (concentrating collectors)
Water in insulated tanks	+	-	-	-
Water in uninsulated tanks with earth around	+	-	+	-
Crushed stone	-	+	+	-
Dry earth	+	+	+	-
Wet earth	+	-	-	-
Wetted earth with dry earth around	+	-	-	-
Deep aquifer	+	-	-	-
Near surface insulated aquifer	+	-	-	-
Eutectic salts	+	+	-	-
Pressured layer plus thermal layer	+	-	+	+

The topic has attracted many theoretical studies, such as the work of Cavalleri and Foligno (65), who suggested the use of a large artificial lake capable of heating a city with a million inhabitants, and Shelton (66) who studied the thermal interaction between any underground storage system and the surrounding ground analytically. Among many ideas which were to lead to practical interseasonal storage systems by the 1980s was the floating 'shallow pond' collector concept suggested by Grevskott (67) in 1977. A thin layer of water was held in a plastic bag on a floating wooden frame under a transparent plastic cover during the day and was allowed to drain into the reservoir or lake on which the collector floated.

A significant confirmation that interseasonal storage had become a practical proposition in a cold northern climate by the start of the 1980s came in a report from Roseen and Perers of Studsvik Energiteknik AB in Sweden (68). Their system is based on cheap excavated and insulated pits, or heat storage tanks, with floating insulating covers. The solar collectors are mounted on the cover, which rotates to track the sun. Integrated units of this type could be connected to a normal district heating network, or to a low temperature heating system. The prototype plant has been in use since February 1979. The volume of the heat store is 640 m^3 and the collector area is 120 m^2. The system is designed to supply the annual heat consumption of a low temperature heated office building with a floor area of 500 m^2. The test programme under way in 1981 included measurements of the energy flow, efficiency, and material degradation. The results from the first two seasons agreed with the design values. The energy collected during 1979 and 1980 was 30.7 and 35.4 MWh, respectively. The corresponding average efficiency of the solar collectors was 42% in 1980 when the system provided 97% of the total heat demand of the building. The azimuth tracking system has been found to increase the energy collected annually by 48% as compared to fixed collectors. Based on the costs of the prototype the calculated cost of a full size plant shows that the necessary investment for a one family house is around $12 000 (at 1981 prices). Of this investment, 51% is for solar collectors, 21% for heat storage, 13% for distribution, and 15% for the heat pump.

Another Swedish project uses deep ground heat storage by penetrating rock or clay to a depth of up to 100 m by vertical holes spaced up to 4 m apart (69). Solar heat collected by very simple unglazed collectors is used to heat water which is circulated deep underground. A single family house with a 10 000 m^3 rock volume as the basic heat store has been successfully tested and the concept promises to be competitive with conventional fuels during the 1980s. A scheme for the interseasonal storage of heat in a cylindrical layer of soil 50 m in diameter and 20 m deep was reported in Holland (70). This store would be linked to a group of at least one hundred houses.

The search for new storage materials continued to bring forward some interesting possibilities in the 1980s. For example, Chabazitic tuff is a zeolitic rock found in Italy and, as shown in Table 4.11, has a considerable potential as a thermal energy store (71).

Table 4.11.

Material	Volumetric heat capacity $\times 10^{-8}$ (J/m^3)
Water	2.51
Granite	1.29
Dry earth	0.59
Chabazitic tuff	2.68

Fig. 4.21.

Fig. 4.22.

Space Cooling Storage

Residential cooling by circulating the hot air extracted from a building
through pipes buried in the ground has been studied in the USA (72, 73).
One conclusion was that 'intuitive' designs appear to grossly overestimate
the cooling available in late summer, as the ground would have been heated.
Large diameter pipes with large air flows are needed for economical
operation.

The House as a Heat-store

The use of the space heated house as the energy store is well known, but an
original idea in conserving heat inside the house is the Zomework's bead
wall, developed by Steven Baer (74). This is illustrated in Fig. 4.21
(emptying) and Fig. 4.22 (filling). Styrofoam beads are automatically blown
between two panes of glass to prevent heat losses at night in the winter
months or they can be used to stop unwanted heat gains in the summer. The
system has an advantage compared with folding doors or shutters which
require free space next to the window.

REFERENCES

(1) Telkes, M., A review of solar house heating, *Heating and Ventilating*,
     46, 68-74, September 1949.

(2) Hottel, H. C. and Woertz, B. B., The performance of flat-plate solar-
     heat collectors, *Trans ASME* 64, 91-104, 1942.

(3) Nemethy, A., Heated by the sun, *American Artisan*, Residential Air
     Conditioning Section, August 1949.

(4) Hutchinson, F. W., The solar house, a full-scale experimental study,
     Heating and Ventilating 42, 96-97, September 1945.

(5) Hutchinson, F. W., The solar house, a research progress report,
     Heating and Ventilating 43, 53-55, March 1946.

(6) Hutchinson, F. W., The solar house, a second research progress report,
     Heating and Ventilating 44, 55-59, March 1947.

(7) Löf, G. O. G., Solar energy utilization for house heating, Office of
     the Publication Board, PB 25375, 1946.

(8) Solar house heater yields 20% fuel saving in University of Colorado
     Experimental Installation, *Arch. Forum* 86, 121, February 1947.

(9) Shurcliff, W. A., Solar heated buildings - a brief survey, 19 Appleton
     St., Cambridge, MA. 02138, USA. 8th edition, March 1975.

(10) Steadman, P., Energy, environment and building. A report to the
     Academy of Natural Sciences, Philadelphia, CUP, 1975.

(11) Hottel, H. C., Residential uses of solar energy, Proc. World Symposium
     on Applied Solar Energy, Phoenix, Arizona, 1955.

(12) Dietz, A. G. H. and Czapek, E. L., Solar heating of houses by vertical wall storage panels, *ASHVE J. Heating, Piping and Air Conditioning* 22, 118, March 1950.

(13) Engebretson, C. D., Use of solar energy for space heating: MIT Solar House No. IV, Proceedings of UN Conference on New Sources of Energy, Rome, 1961, pub. United Nations, New York, 1964.

(14) Engebretson, C. D. and Ashar, N. G., Progress in space heating with solar energy, Paper number 60-WA-88, Winter ASME Meeting, November 27 to December 2, 1960.

(15) Hottel, H. C. *et al.*, Panel on solar house heating, Proceedings of the World Symposium on Applied Solar Energy, Phoenix, Arizona, 1955.

(16) Bridgers, F. H., Paxton, D. D. and Haines, R. W., Performance of a solar heated office building, *Heating, Piping and Air Conditioning* 27, 165-170, November 1957.

(17) Gilman, S. F., Evaluation of a solar energy heat pump system, ISES Congress, Los Angeles, Extended Abstracts, Paper 42/8, July 1975.

(18) Thomason, H. E., Solar space heating and air conditioning in the Thomason house, *Solar Energy* 4, 11-19, 1960.

(19) Thomason, H. E., Solar-heated house uses $\frac{3}{4}$ hp for air conditioning, *ASHRAE J.* 4, 56-62, 1962.

(20) Thomason, H. E., Experience with solar houses, *Solar Energy* 10, 17-22, 1966.

(21) Thomason, H. E. and Thomason, H. J. L., Solar houses - heating and cooling progress, *Solar Energy* 15, 27-39, 1973.

(22) Böer, K. W., The solar house and its portent, *Chem Tech* 3, 394-399, July 1973.

(23) Böer, K. W., Higging, J. H. and O'Connor, J. K., Solar One, Two Years Experience, Institute of Energy Conversion, University of Delaware. (Presented as paper 42/3, ISES World Congress, Los Angeles 1975, jointly with Kuzay, T. M., Malik, M. A. S., Telkes, M. and Windawi, H. M.)

(24) Copper Development Association, 405 Lexington Avenue, New York, Press release, July 1975.

(25) Reynolds, J. S., Larson, M. B., Baker, M. S., Mathew, H. and Gray, R. L., The Atypical Mathew solar house at Coos Bay, Oregon, ISES Congress, Los Angeles, Extended Abstracts, Paper 42/13, July 1975, and in *Solar Energy* 19, 219-232, 1977.

(26) McDaniels, D. K., Lowndes, D. H., Mathew, H., Reynolds, J. S. and Gray, R. L., Enhanced solar collection using reflector-solar thermal collector combinations, ISES Congress, Los Angeles, Extended Abstacts, Paper 34/11, July 1975.

(27) Mathew, H., Private communication, 1975.

(28)   Kreider, J. F., The Stationary Reflector/Tracking Absorber Solar
       Concentrator, US ISES Annual Meeting, Fort Collins, Colorado, 1974.

(29)   Pike, A., The Cambridge Autonomous House, UK ISES Conference on Solar
       Energy in Architecture and Planning, April 1975.

(30)   Energy and housing Symposium, Open University, Milton Keynes, *Building
       Science*, 127, 31 October 1974.

(31)   Szokolay, S. V., Design of an experimental solar heated house at Milton
       Keynes, UK ISES Conference on Low Temperature Collection of Solar
       Energy, April 1974.

(32)   Solar heated house in Milton Keynes, Milton Keynes Development Corp.,
       Wavendon Tower, Wavendon, Milton Keynes MK17 8LX, UK.

(33)   Maw, R., Boyd, D. and Thring, J. B., Design and economic analysis of
       solar heating systems integral with housing, *Solar World Forum*, Vol. 1,
       355-360. Pergamon Press, Oxford, 1982.

(34)   McLaughlin, T. P., *A House for the Future*, Independent Television Books
       Ltd, London, 1976.

(35)   Wilson, D. R., Solar collection and storage techniques in a house
       conversion at Macclesfield for Granada TV, UK ISES Conference (C13) on
       Practical aspects of domestic solar water heaters, London, October
       1977.

(36)   Stephen, F. R., Performance monitoring of low energy house,
       Macclesfield, *International Journal of Ambient Energy*, 1, 29-46, 1980.

(37)   Seymour-Walker, K., Low energy experimental houses, *BRE News* 34,
       12-13, 1975.

(38)   Bruno, R., Hermann, W., Horster, H., Kersten, R. and Madhjuri, F., The
       utilisation of solar energy and energy conservation in the Philips
       Experimental House, ISES Congress, Los Angeles, Extended Abstracts,
       Paper 41/8, July 1975.

(39)   Philips Forschungslaboratorium, Aachen, The Experimental House, 1975.

(40)   Bourne, R. C., A volume collector-heat pump demonstration house, ISES
       Congress, Los Angeles, Extended Abstracts, Paper 14/13, July 1975.

(41)   Wormser, E. M., Design, performance and architectural integration of a
       solar heating system using a reflective pyramid optical condenser, ISES
       Congress, Los Angeles, Extended Abstracts, Paper 52/8, July 1975.

(42)   McVeigh, J. C., Solar heating feasibility report, Appendix D, Clumber
       Park: an interpretive study, Countryside Commission, Cheltenham, 1976.

(43)   RWE informiert 166, Energie-Dach, Energie-Fassade, Energie-Zaun,
       Energie-Stapel. Rheinisch-Westfälisches Elektrizitätswerk AG, Abt.
       Anwendungstechnik, Essen, 1981.

(44)   *ISES News* No. 12, 7 June 1975.

(45) Besant, R. W. and Dumont, R. S., Comparison of 100 percent solar
      heated residences using active solar collection systems, *Solar Energy*
      22, 451-453, 1979.

(46) Turrent, D., Performance monitoring of solar heating systems in
      dwellings, *Solar World Forum*, Vol. 1, 820-826. Pergamon Press, Oxford,
      1982.

(47) University of Wisconsin-Madison, Engineering Experiment Station Report
      38-9 "TRNSYS - A Transient Simulation Program" and Report 38-10
      "TRNSYS - A Transient System Simulation User's Manual" 1978-9.

(48) Persons, R. W., Duffie, J. A. and Mitchell, J. W., Comparison of
      measured and predicted rock bed storage performance, *Solar Energy* 24,
      199-201, 1980.

(49) Braun, J. E., Klein, S. A. and Mitchell, J. W., Seasonal storage of
      energy in solar heating, *Solar Energy* 26, 403-411, 1981.

(50) Evans, D. L., Rule, T. T. and Wood, B. D., A new look at long term
      collector performance and utilizability, *Solar World Forum*, Vol. 3,
      2523-2529, Pergamon Press, Oxford, 1982.

(51) Jesch, L. F., Greeves, T. W. and Haralambopoulos, D. A., Simulation of
      optimal control strategies for large water pre-heat systems, *Solar
      World Forum*, Vol. 1, 119-123. Pergamon Press, Oxford, 1982.

(52) Keyes, J. H., Project Sungazer: A vertical-vaned flat plate collector
      with forced-air heat transfer, Proc. Workshop on Solar Collectors for
      Heating and Cooling of Buildings, 21-23 November 1974, NSF-RANN-75-019
      May 1975.

(53) Close, D. J. *et al.*, Design and performance of a thermal storage air
      conditioning system, Mechanical and Chemical Transactions of the
      Institute of Engineers, Australia, 4, 1968.

(54) Read, W. R., Choda, A. and Copper, P. I., A solar timber kiln, *Solar
      Energy* 15, 309-316, 1974.

(55) Dunkle, R. V., Design considerations and performance predictions for an
      integrated solar air heater and gravel bed thermal store in a
      dwelling, Australian and New Zealand section ISES, Melbourne, July
      1975.

(56) Close, D. J. and Pryor, T. L., The behaviour of adsorbent energy
      storage beds, ISES Congress, Los Angeles, Extended Abstracts, Paper
      31/2, July 1975, and in *Solar Energy* 18, 287-292, 1976.

(57) Telkes, M., Solar energy storage, *ASHRAE Journal*, 38-44, September
      1974.

(58) Page, J. K. R., Swayne, R. E. H., Mead, I. K. and Haymen, C., Thermal
      storage materials and components for solar heating, *Solar World Forum*,
      Vol. 1, 723-730, Pergamon Press, 1982.

(59) Johnson, T. E., Lightweight thermal storage for solar heated buildings,
      Solar Energy 19, 669-675, 1977.

(60) Busic, V., Carfagna, C., Salerno, V., Vacatello, M. and Fittipaldi, F., The layer perovskites as thermal energy storage systems, *Solar Energy* 24, 575-579, 1980.

(61) Severson, A. M. and Smith, G. A., Salt thermal energy storage for solar systems, ISES Congress, Extended Abstracts, Paper 31/5, July 1975.

(62) Kosaka, M. and Asakina, M., Discussions on heat storage material at low temperature level, Paper 31/3, *Ibid.*

(63) Ervin, G., Solar heat storage based on inorganic chemical reactions, Paper 31/6, *Ibid.*

(64) Givoni, B., Underground long-term storage of solar energy - an overview, *Solar Energy* 19, 617-623, 1977.

(65) Cavalleri, G. and Foligno, G., Proposal for the production and seasonal storage of hot water to heat a city, *Solar Energy* 19, 677-683, 1977.

(66) Shelton, J., Underground storage of heat in solar heating systems, *Solar Energy* 17, 137-143, 1975.

(67) Grevskott, G., The use of water reservoirs, lakes and bays for shallow pond collectors and submerged hot water storage, *Solar Energy* 19, 777-778, 1977.

(68) Roseen, R. A. and Perers, B. O., Integral seasonal storage solar heating plant, *Solar World Forum*, Vol. 1, 697-702. Pergamon Press, Oxford, 1982.

(69) Platell, O. B., The sunstore - deep ground heat storage, low temperature collectors and indoor heaters, *Solar World Forum*, Vol. 1, 697-702. Pergamon Press, Oxford, 1982.

(70) van den Brink, G. J. and Hoogendoorn, C. J., Seasonal heat storage efficiency in water saturated soils, *Solar World Forum*, Vol. 1, 760-764. Pergamon Press, Oxford, 1982.

(71) Scarmozzino, R., Aiello, R. and Santucci, A., Chabazitic tuff for thermal storage, *Solar Energy* 24, 415-416, 1980.

(72) Shelton, S. V., Design analysis for direct ground cooling, *Solar World Forum*, Vol. 3, 2099. Pergamon Press, Oxford, 1982.

(73) Sinha, R. R., Goswami, D. Y. and Klett, D. E., Theoretical and experimental analysis of cooling technique using underground air pipe, *Solar World Forum*, Vol. 3, 2105-2114. Pergamon Press, Oxford, 1982.

(74) Harrison, D., Beadwalls, *Solar Energy* 17, 317-319, 1975.

CHAPTER 5

PASSIVE HEATING AND COOLING

Passive solar design uses solar energy naturally by involving the conventional building elements for solar energy collection, storage and distribution. Unlike the active systems in which a carefully designed and relatively complex solar collector is connected to fans or pumps, storage or heat exchange units to provide heating, the passive system uses natural convection, conduction and radiation. More recently, some passive systems have been designed with an active element such as a fan, to promote the circulation of air. In other systems an active domestic water heating system may be combined with a passive space heating system. These systems are generally known as hybrid and will also be included in this chapter. The growth of interest in passive systems has been a very marked feature of recent solar heating applications. This is reflected both by a rapidly expanding literature, in which pride of place must go to the American Mazria (1), and in conferences attracting audiences of thousands in the USA.

The various different approaches to passive solar energy collection were categorized into five basic groups in 1977 by Stromberg and Woodall (2) as follows:

(1) The direct gain approach. This was the system tried in the 1930s in the USA in which there is an expanse of glass facing the equator. The building should ideally have considerable thermal mass, such as a concrete floor or heavy masonry construction. The basic principles go back thousands of years as shown in Chapter 1.

(2) The Trombe wall or thermal storage wall, in which the heat is stored in a wall which also absorbs the solar energy when it comes through the glazing.

(3) The solar greenhouse, in which the features of the solar thermal storage wall and the direct gain approach can be combined by building a greenhouse on the south side of a building (in the northern hemisphere). This approach was attracting very considerable attention in the early 1980s.

(4) The roof pond. In this system a shallow pond or tank of water sits
     on a flat roof with its surface generally contained by a transparent
     plastic sheet. Movable insulation is essential because solar input
     during the summer months can be very large and it is needed to
     protect the building from heat losses in the winter. In the winter,
     the movable insulation covers the top of the pond at night and the
     heat from the pond can radiate down to the house. During the summer
     the insulation covers the top of the pond in the day and is removed
     at night when the heated water can radiate out to the colder night
     sky.
(5) The natural convective loop. In this system air is usually used as
     the heat transport medium and works in the same way as the classic
     thermosyphon water heating system.

Some of the more recent developments, such as the double envelope system,
will also be discussed. The final example draws together in a single design
most of the desirable characteristics of the best known systems.

The Direct Gain Approach - the Curtis House (3, 4) and North Several (5)

The first solar house in the UK, illustrated in Fig. 5.1, was designed by
the architect-owner, the late Edward J. Curtis, who lived in it during his
lifetime. The house was built in 1956 at Rickmansworth, near London, and
was the result of a study which he had carried out during the previous years
into domestic design and environmental control systems from simple heating
devices to total air conditioning. In the design stage a fundamental
principle was that control of the internal environment should be obtained by
a combination of solar energy and a heat pump system to provide heat and
cooling, together with hot water supply and refrigeration. Curtis and others
who have attempted to apply solar principles to space heating in the UK have
appreciated that without a very large, heavily insulated heat storage system
it is only possible to obtain a certain percentage of the total space heating
requirements from solar energy. The overall design aim was to provide a
comfortable and flexible internal requirement to offset the external
temperature conditions of a year round basis. This requirement was also
linked to the aesthetic interpretation of interior spaces where it was
desired to give the maximum feeling of spaciousness with a ground plan of
only 11.2 × 6.1 m.

The site, which is on the top of high ground overlooking a valley, was
selected to give the required orientation and to provide an unobstructed sun
track. The main spaces were planned on the south and west sides while the
east side took the entrance area, landing and two bedrooms. The construction
principle adopted was that of two free standing side walls of brick cavity
construction ends and one large panel in-filled with purpose-made timber
window walling. It was decided to provide maximum glass area on the south
front to exploit the solar gain and with one small exception the entire south
elevation consists of Plyglass clear cavity double glazing units set in
timber frames. Panels to west and north are also double glazed. Year round
air conditioning is provided by a heat pump which originally used air but
was subsequently modified to use water as the low temperature heat source.
Average temperature levels were between 20.6°C for daytime temperatures and
22.0°C for evening periods. The build-up of heat was found to be very
rapid and except for very cold weather the heat pump unit was switched off
at approximately 23.00 h. The high insulation factor enabled the heat
content to be retained until approximately 05.00 h. when the unit commenced
operation giving a comfortable temperature at ground level in the region of
19°C by 07.00 h. The main lessons to be drawn from this particular work are

that the use of large glazed areas to obtain maximum solar gains within the
interior of a house in the UK can substantially reduce the load on heating
appliances, including a heat pump system, but on the other hand these large
glazed areas constitute a high heat loss on dull cold days or evening
periods and during winter nights. Some means of controlling the glazed
areas must be installed and used efficiently to conserve solar build-up -
even to the point of reducing daylight penetration to the absolute minimum.
In the Curtis house it is possible to have 80% of the glazing covered by
heavy curtaining with the remaining 20% giving sufficient daylight for the
interior. The total heating/cooling/solar system operated very
satisfactorily for the twenty-year period in which the experience was
reported, with overall annual running costs approximately one-third of those
of similar houses in the same neighbourhood.

Fig. 5.1. The Curtis house.

The use of direct solar gain also featured prominently in North Several,
Blackheath, London, a block of seven town houses designed by the architect
Royston Summers, who has lived in one of them since it was built in 1968.
His aim was ". . . to design a building in which all elements - viewed as
contributions to an integrated totality - would be used or would perform at
their optima, all factors would interrelate to the maximum relevant degree,
and all physiological and psychological 'desiderata' would be met . . .".
To achieve this he adopted an overall conservation strategy. For example,
he minimized the quantity of materials used and ended with a total
construction cost which was less, per unit of floor area, than that of
local authority housing.

Solar energy was exploited both by direct gain through the large south-facing
double-glazed windows and through the use of a 'thermal trap' by the
entrance to the front-door of each house. As shown in Fig. 5.2 this is an
external enclosure in which air movement is limited, with dark-coloured
surfaces and made with materials having a high thermal capacity. The
overall contribution which direct solar gain makes to the total annual space-
heating requirements is between 15 and 20%. Summers admits that the north
side, which also has a floor-to-ceiling double-glazed wall is ". . . a
thermal embarrassment in winter . . ." but there are compensations in
'perceptual' terms with a fine view over the open spaces of Blackheath
common.

Fig. 5.2.
North Several, Blackheath.

Fig. 5.2.

St George's School, Wallasey (6-8)

Probably the best-known solar building in Europe, the annexe to St George's
School, Wallasey, was designed by the late A. E. Morgan and built in 1962.
It contains a large solar wall and its ability to maintain good thermal
conditions during the winter months without any conventional form of central
heating has attracted considerable attention. The Department of the
Environment have sponsored studies on the use and thermal response of the
annexe which have been carried out under the direction of Dr M. G. Davies of
the University of Liverpool.

The main solar wall occupies the entire south-facing wall of the building
and is 70 m long × 8.2 m high. A mean U-value of 3.1 W/m^2 K was assumed for
all heat balance calculations. Most of the wall is double-glazed with 600 mm
separation between the leaves. Each classroom, however, is provided with
two or three opening windows, which constitutes areas of single-glazing.
The response time of a room in the building is about six days for a zero air
change, falling to three days with two air changes per hour. The depth of
the building from south to north is approximately 11.5 m. The ground floor
consists of 100 mm of screed upon 150 mm of concrete. The intermediate floor
is of concrete approximately 230 mm thick and the roof consists of
approximately 180 mm concrete slab with a layer of 126 mm of expanded
polystyrene above it, suitably protected. The partition walls are of 230 mm
plastered work. On the north side at first-floor level the external walls
are also of 230 mm brick with the 125 mm polystyrene external cladding. The
mean U-value here is 0.24 W/m^2 K. At ground-floor level the external wall
is partly ranch walling and partly solar wall similar to the south side.
Overall, the U-value for the building is 1.1 W/m^2 K. The only sources of
heat normally operating in a typical classroom in the annexe are due to the
occupants, the electric lighting and to solar radiation.

The older part of the school houses a similar number of students, about 300,
and the two parts serve similar functions so that one acts as a control on
the other. Both staff and students in winter preferred the thermal
environment of the annexe to that of the main school, which was often
bitterly cold. However, the low ventilation rate which was necessary to
maintain the temperature in the annexe sometimes allowed the atmosphere to
become rather unpleasant. A further minor problem was noise in the summer.

At the time when the building was erected the authorities insisted that a
conventional central heating system should be installed in case the solar
wall did not work. It was not until towards the end of the winter in 1973
that the central heating system was needed for the first time, but only
because vandals had broken some of the windows in the solar wall. Based on
unit costs for heating other schools in the same district, it is estimated
that about 30% of the total heating requirements of the annexe are provided
by solar energy.

The Trombe Wall
The French solar housing research programme started in 1956 at the Centre
National de Recherche Scientifique (CNRS) when the Trombe wall principle was
first patented (9-11). It is interesting to note the similarity between
this system and that described by Professor Morse a hundred years ago (12).
The basic principle is that thermally massive south-facing walls, usually
made of concrete, are painted black, or some other relatively heat absorbing

colour such as red, dark green, or dark blue, and covered with glass on the
outside, leaving an air gap between the wall and the glass. The wall is
both a heat collector and a heat store. As solar radiation passes through
the glass it is absorbed by the surface coating which heats the wall. As
the long wave re-radiation from the wall is trapped behind the glass, the
air between the wall and the glass becomes heated. Ducts at the top and
bottom of the wall allow the heated air to be fed into the room at ceiling
level, while the colder air from the floor is drawn in at the bottom, as
shown in Fig. 5.3. For summer cooling, the valves at the top of the wall
are arranged to vent the heated air to atmosphere and the valve at the rear
of the building, at ceiling level, is opened to allow cooler air to flow
through. The walls are typically 300-400 mm thick. It would be possible
to have other heat storage systems within the walls, such as water tanks or
change-of-state chemical storage. The prototypes at Odeillo were
aesthetically unattractive because they were poorly insulated and had very
small south facing windows. In the latest designs the ratio of collector
area to volume of the house is 0.1 m^2 per m^3 and it is now quite difficult
to distinguish between the solar collectors and the windows. The latest
house has been described as having the general appearance of a classical
dwelling.

The French estimate that between 60 and 70% of the heating load in
Mediterranean climates, such as at Odeillo, and between 35 and 50% in less
favourable climates, such as Chauvency le Chateau, Meuse, can be provided by
this system. The main lessons which can already be drawn are as follows:

(i)   there are no problems of mechanical resistance to flow as with
       conventional roof-mounted water heaters;
(ii)   there are no leakage problems; and
(iii) there are no problems associated with freezing.

The latter consideration is probably most significant for the UK, but
although in temperate northern European countries a slightly greater
percentage of the available solar energy lands on a vertical south facing
surface in mid-winter, the scarcity of direct radiation or sunny winter days
is a distinct disadvantage.

A Trombe Wall System in the USA - The Kelbaugh House (1, 2, 13, 14)

This house in Pine Street, Princeton, New Jersey, was designed and built by
the architect and solar consultant, Doug Kelbaugh, who lives in it with his
family. The house, shown in Fig. 5.4, is situated along the northern
boundary of the site, which allows unshaded access to the maximum available
sunlight, preserved a large tree and gave a large single outdoor space. The
very marked contrast between the house and its neighbours shows a different
approach to planning permission in this case from many suburban situations
in the UK, for example, where such flexibility would be unlikely.
Kelbaugh's aim was to design a house with a maximum construction cost of
$45 000 (1974 prices), a Trombe wall, one large interior room (living/
reception), a fireplace, three bedrooms, an attached solar greenhouse and a
large exterior garden area (or yard as it is known in the USA). The ground
floor is one large continuous room, with the living room and dining room
area at the western end partially separated from the kitchen by the stairs.
The solar greenhouse is immediately to the south of the kitchen and
approached from within through a large arch. Upstairs the bathroom and
other plumbing fixtures are above the kitchen to reduce long pipe runs, and
the three bedrooms are in line along the upper part of the Trombe wall.

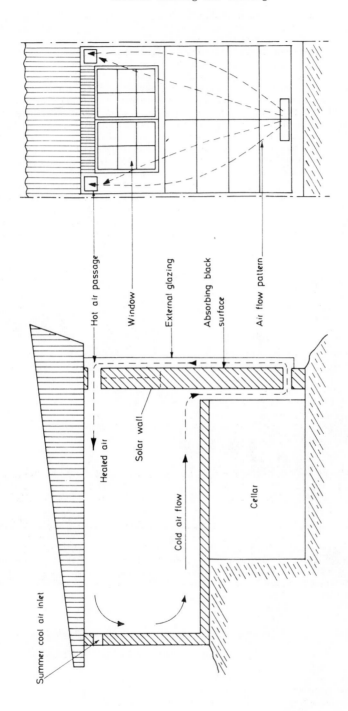

Fig. 5.3. Trombe three-unit dwelling.

Cost limitations and a desire to make the air circulation easier to control
and predict kept the room height at normal single floor height throughout.
Window area was kept to a minimum, particularly on the north wall. The
three non-solar walls have standard wood frames with rough-sawn cedar
plywood on the exterior. The cavities – 110 mm on the ground floor, 90 mm
for the first floor and 240 mm for the roof above the first floor ceiling –
were filled with cellulosic fibre (recycled newspaper) to achieve an R factor
of 3.2–3.5 m^2 $^{\circ}C/W$ in the walls and 7.0 in the roof. The outside of the
foundation wall around the perimeter was protected by 25 mm of styrofoam to
a depth of 600 mm.

The Trombe wall was constructed of concrete, 400 mm thick and painted with
a selective black paint over a masonry sealer. The glazing placed in front
of the wall consisted of two sheets of double-strength window glass. Vents
in the wall at the top and bottom of each floor level completed the natural
convection loop. During the hot summer days, four fans at the eaves of the
Trombe wall assist the flow of cooler air through the house from the
northern windows.

Fig. 5.4.   The Kelbaugh House.

The back-up heating system consists of a conventional gas-fired hot-air
furnace with a net delivery of about 60 MJ/h. The wood-burning fireplace
which is situated in the centre of the Trombe wall can provide additional
heat. It also alters the air circulation pattern by drawing in cooler
outside air and its contribution to the space heating requirement was
neglected.

Table 5.1. General data

Trombe wall - vertical area excluding the greenhouse but including six direct gain windows	56 m^2
Greenhouse - projected vertical area	17.5 m^2
floor area	20 m^2
House - heated floor area	195 m^2
heated volume	538 m^3
Total construction costs, excluding owner's labour (1975 prices)	\$55 000

Performance

Princeton is located at a latitude of 40° 21'N and averages 2830°C-days.
Unlike the UK, it receives about 55% of possible direct sun in the winter
months. The results from the first two winters of operating experience are
shown in Table 5.2. The second winter, 1976-1977, was more severe, but
modifications to the system substantially improved the performance.

Table 5.2. Performance for the first two winters

	1975-1976 2500°C-days	1976-1977 3000°C-days
Heat loss (MJ)	130 950	136 736
Miscellaneous heat gains* (MJ)	17 830	17 830
Net space heating requirement (heat loss) - (heat gains) (MJ)	113 120	118 916
Gas furnace (at 75% efficiency) (MJ)	27 320	18 991
Solar contribution to net space heating requirement (MJ)	85 800	99 925
Percentage	76%	84%

*Includes gains from cooking, lighting, domestic hot water and
occupancy.

The indoor temperatures were allowed to swing up to $3^{\circ}C$ during the 24-hour
cycle, allowing the Trombe wall to collect and discharge heat. The
seasonal low and high temperatures were 14 and $20^{\circ}C$ downstairs and $17-22^{\circ}C$
upstairs. Several modifications were made to the system after the first
winter. Reverse thermosyphoning down the Trombe wall surface was
eliminated by installing a passive damper on the inside of the upper air
vents - originally a screen fitted with a light cloth - but thin polythene
sheeting would be equally effective. A door at the top of the stairs has
greatly reduced the heat flow to the upper floor. Excessive temperature
fluctuations were observed in the greenhouse on sunny winter's days,
typically from $24-14^{\circ}C$ over a few hours after sunset. This problem was
solved by providing extra thermal mass - eight five-gallon water drums
painted black. The original heat losses from the greenhouse represented
over 40% of the total losses for the whole house and these were greatly
reduced by double-glazing.

Among the advantages which Kelbaugh listed for the Trombe system were its
long life, comfortably low operating temperatures, architectural integration
into the building, lack of moving parts and its simplicity. He felt that
having a system which could be easily understood was very important:
"The Trombe system is refreshingly simple and does not depend on all of its
parts working. It is easily understood by most people that I have explained
it to, which is an eminently important aspect in the public acceptance of it
or any other solar heating system."

The Solar Skin or Single-Glazed Wall

If most of today's buildings could be fitted with passive solar systems over
the next twenty years, the reduction in the world's energy demand could be
considerable. One of the major obstacles is the lack of information on the
most simple and cost effective means of achieving some passive gains. An
approach which has been successful in Scotland (15) is to apply a vertical
glazed panel to the external walls of a building. No connecting passages or
ducting to the inside of the building is necessary. The glazed panel acts
both as a weather resistant skin and additional external insulation, with
the result that the solar gain through the wall is increased while the basic
heat loss through the wall is reduced. This work is particularly relevant
within the British Isles because as latitude increases so the cost
effectiveness of solar space heating using vertical collectors also
increases due to the longer heating season (16).

A somewhat similar approach to retrofitting has been tried in Canada (17)
where the existing forced ventilation system in a building has been used to
draw air through the glazed skin.

The Transwall (18, 19)

The concept of a partially transparent solar wall placed adjacent to a
window admitting solar energy was first suggested by McClelland and Fuchs
(18). Part of the solar energy is absorbed within the transwall and the
remaining part is transmitted to the interior of the room. This system
combines the features of the direct gain approach and the Trombe wall, and
is considered to be architecturally more attractive than the latter as it
admits light to the room and allows the occupants to see out through the
window without the possible overheating or glare problems of the direct gain
window. After three years of development a 'second generation' system was

assembled in 1981 with conventional commercially available glass and
aluminium components (19) using established manufacturing methods for low
volume production. Each water-containing module has overall dimensions of
0.61 m (horizontal) by 0.406 m (vertical) with a 0.185 m in thickness. Light
transmission was approximately 30%. The thickness and transmission
parameters were optimised by a computer model validated by test results from
a first generation prototype.

The 'Double Envelope' Concept for Heating and Cooling

The principle extends the basic Trombe wall concept by separating the
collector from the thermal storage mass. Natural convection from a south
facing (in the northern hemisphere) glazed wall carries the heated air
upwards and through the space in a double skinned roof. This approach would
be particularly desirable in the UK where a rapid response to the available
solar energy is important. The movement of the air is controlled by the use
of small fans and it can pass down through the north wall to an underground
thermal rock storage bin. Storage periods of several days may be achieved
without significant increases in building costs (20) and this, in turn, can
lead to solar heating providing some 60-80% of the total heat demand.

The double envelope roof can be used in lower latitudes for one or two
storey buildings to provide a simple solar air conditioning system. A
Japanese system (21) consists of a corrugated and glazed copper roof with an
air gap underneath separating the copper from a second, flat roof. The air
in this gap can be vented to the exterior or passed to the interior of the
house through a series of ducts.

In the summer, water trickles down the copper roof and evaporates,
dissipating the excess heat from solar radiation. At the same time, outside
air is passed through the gap in the double roof and vented to the atmosphere.
This effectively prevents any heat from the outer surface reaching the
interior of the house and enables comfortable conditions to be maintained
with a relatively small electrically powered air conditioner. Night-cooling
is also possible by a combination of radiative and evaporative cooling, and
directing the cooler air flow from the gap into the house. Preliminary
results show a reduction of up to 50% in the cooling load in the summer
months.

In the winter, the gap can act as an additional sealed thermal insulating
layer. When the roof reaches higher temperatures and this air is above the
house temperature, it can be passed through the ducts to heat the house.

The Roof-pond Passive System (Hay House) (22-25)

The use of roof ponds for cooling buildings was established many years ago,
but it was not until 1967 that Hay developed a system using water enclosed
in black polyvinylchloride bags to form the roof pond. In the prototype
test building at Phoenix, the water was held in a layer about 180 mm deep
which acted both as a heat collector and storage medium. The bags were
supported on a flat metal roof, which also acted as a heat-exchanger and
ceiling for the building. At night, insulated panels were placed over the
bags to prevent heat loss. During the summer, the procedure was reversed,
so that nocturnal radiation to the sky could take place, leaving a cooler
pond in the morning which was then insulated by the panels, thus providing
natural cooling to the building during the day.

The system was subsequently tested on a major two-year evaluation project
carried out by California Polytechnic State University on a three bedroom,
two bathroom, single-storey house at Atascadero, California. The living
area of the house, about 106 m², was slightly larger than the roof pond area.
The movable insulation panels were electrically powered and controlled
manually or by a differential temperature controller. The following extract
from the report (26) on the tests indicate how successfully the building
performed:

> The thermal performance of the house was very positive. The moving
> insulation system supplied 100% of the heating and cooling requirements
> of the building during the test months. During this time, the system
> was able to keep the indoor temperature between the extremes of 19°C and
> 23.3°C except during special test periods or times of prototype
> breakdown. Even during these exceptional periods, the temperature never
> got higher than 26°C or lower than 17°C . . .

Temperatures as high as 38°C were experienced in July and the lowest
recorded temperature was -3°C in February 1974, when the monthly mean
average daily outdoor temperature was 8.3°C. The use of an inflatable
plastic cover enabled the system to be operated in both an unglazed and
single-glazed configuration. During the summer it was essential to keep the
cover deflated so that the living space could be maintained at less than
27°C.

Haggard (25) believes that information from the project can be used to
investigate architectural extensions of the system which would be capable
of handling other climates and spatial needs. These include multi-storey
developments with movable insulation in south-facing wall channels and
folding insulation panels acting as snow shields on flat roofs.

Cooling

There are now a number of excellent examples of passively cooled buildings
in which careful attention to the national and local climatic restraints
combined with local traditions in building structure and materials, produce
a cool and comfortable internal environment. The Samuel Jackson Prescod
Polytechnic in Barbados, completed in 1981, was reported to meet these
requirements (27) by using the following techniques:

   (i)   grouping the buildings with grassed acres and trees to encourage
         breeze and reduce glare and noise disturbance;
  (ii)   controlling breezes, internally and externally, by window and
         valance design; and
 (iii)   careful attention to roof design to facilitate the release of heated
         air by convection and the elimination of radiant heating by the use
         of a sound insulating membrane.

Some Passive Projects in the UK

The announcement in February 1975 of a project to build nine solar heated
houses at Higher Bebington near Liverpool, attracted considerable interest
as the planning permission had been granted by the Wirral Urban District
Council who had been involved in the approvals for the St George's School
annexe some fifteen years earlier. The scheme was the result of a joint
approach from the glass manufacturers, Pilkington Brothers, and the

Loughborough University of Technology and was financed by the Department of the Environment through a local non-profit making housing association.

Acorn Close is a development of 14 houses built of high density brick and all now insulated to a much higher level than the 1976 UK building regulations. Nine of the houses have a south-facing outer wall covered with double-glazing to absorb solar radiation - the basic Trombe wall approach. The other five houses have the same plan form and are used for comparison (28). All nine houses have independent fan-assisted cooling systems to augment the natural ventilation and reduce the effect of excessive solar gain in the summer.

Estimated energy savings of between 30 and 60% were predicted. During construction 25 temperature sensors were incorporated into the structure of the houses. These are connected to a data logger on site linked to the Pilkington Laboratories computer some 15 km away. The energy used in all 14 houses is recorded weekly. The results for the period November 1978 to July 1980 confirm the estimated savings, with the average space heating requirements of five of the solar houses being some 35-45% less than the control group of conventional houses (29, 30).

A passive project believed to be the largest monitored group of direct gain houses in the world in 1982 is the Pennyland Demonstration Project at Milton Keynes. An estate of 177 houses has been laid out to minimize overshading. The estate is divided into two sections having different insulation levels. A further sub-division involves south-facing houses with the glazing split equally between the north and south faces, with the other houses having their main living areas and glazing concentrated on the south side. Both groups are being compared with houses at different orientations and also with a group of houses randomly orientated and overshaded on a nearby estate (31). Preliminary results are quite promising, with savings in the order of 30% being common.

A group of eight larger direct-gain passive houses have been built at Great Linford - some 800 m from Pennyland. The houses have a large area of south-facing double glazing and very small windows on the north side. A very elaborate monitoring system has been devised for the seven occupied houses and the one which is being used as a research test house (32). Slightly better results than Pennyland are predicted.

Physical and mathematical modelling of solar walls at Leeds University has resulted in the development of a design which allows heat recovery by forced convection through ducts within the wall. The wall is treated simply as an air heating element in a building and direct heat transfer at the inside surface of the wall is made very small by insulation. The double-glazing for the solar wall completely seals the space between the absorbing wall surface and the inner sheet of glass (33).

Some Passive/Low-energy Houses in Canada

Commercial cost-effective low energy houses, some with Trombe walls and all with direct gain windows, began to appear on the Canadian market in 1980 (34). A group of 14 were built in Saskatoon, Saskatchewan, by 13 different contracting companies on one south-facing city block. Saskatoon, latitude 52° 08'N, has a continental climate with an average of 6039°C-days of space heating per year and 1003 bright sunshine hours during the seven-month heating season between October and April inclusive. Each house was designed

to consume less than 40 GJ of natural gas fuel energy per year for
auxiliary space heating. This is at the most one third of the energy demand
by conventional houses in the region.

The energy saving features included air-tight vapour barriers with natural
ventilation rates of less than one-tenth of a volume change per hour,
insulation levels that vary between 5.3 and 10.6 $m^2{}^\circ C$ W^{-1} in the walls and
8.8 and 10.6 $m^2{}^\circ C$ W^{-1} in the ceiling, heat recovery from ventilation exhaust
air using air to air heat exchangers with temperature recovery ratio between
0.5 and 0.8 and south-facing window areas from 5.8 to 19.4 m^2 or 59-85% of
the total window area. Window glazings vary from two or four, some with
night shutters and two with Trombe walls.

An interesting feature of the monitoring programme was that it would be
supplemented by health related tests to check on the radon gas levels.

A Hybrid Solar House, Tokyo (35)

Professor Ken-ichi Kimura, who reported on the design and operating results
from this house in 1981, has defined the hybrid solar house as one "in which
energy flows by passive and active solar systems simultaneously and
cooperatively". After the house had been built he observed that no clear
distinction between the passive and the active systems could be found.

The house is built in the suburbs of Tokyo ($2000^\circ C$-days based on $18^\circ C$) and
was occupied by two adults and three children. Insulation for the concrete
walls and the ceilings helps to store the direct gain solar radiation from
double-glazed south windows. These are fitted between the glazing with
venetian blinds having reflective slats. Flat plate collectors with a total
area of 24 m^2 are built into the roof. The collectors have copper pipes with
aluminium fins and a selective surface. The total floor area is 119.51 m^2.
Water is circulated from the collectors to the water storage tank, from which
hot water is supplied through a heat exchanger to the taps. Warm water is
supplied to the pipes in an underfloor space heating system. Comfort was
maintained in the living area by keeping the temperature between $18-22^\circ C$ in
winter without auxiliary space heating. The domestic hot water was sometimes
heated by an auxiliary heater.

The basic design concepts were as follows:

(i)   to design a hybrid solar system with no auxiliary energy source
       required for space heating;
(ii)   to achieve a low annual energy consumption; and
(iii) to use only electricity as the conventional energy source for reasons
       of safety and cleanliness.

The order of priority in energy planning then emerged:

(i)   have a high degree of insulation;
(ii)   use as much direct solar gain as possible;
(iii) use electrical power to move energy around the building, thus making
       more effective use of solar gains;
(iv)   provide solar collectors and a storage tank for the active system; and
(v)   minimize the use of conventional energy.

As all the individual rooms face south-east or south-west, with the
exception of the ground floor living-room, which is situated under the solar
roof and faces due south, the direct gain solar radiation is considerable.
In winter the reflective slats direct the sunlight upwards towards the heavy
concrete ceilings and walls. The equivalent thermal mass is 40 m^3 of water.
The storage tank volume is 950 litres and, in the hot season, excessively
hot water can be circulated to pipes in the earth under the living room.
This acts as a long-term heat store. Kimura pointed out that all the main
objectives were achieved, with a particular success being the storage of
the direct gain solar radiation after it had been reflected from the
venetian blinds. The total annual electric power consumption of the house
from May 1979 to April 1980 was only 5028 kWh. No other conventional energy
source was used.

What do we Measure and Control?

An important characteristic of any passively heated building is the effect
the building has on the comfort of the occupants. Most design procedures
assess comfort conditions in terms of air temperature and other
environmental and physiological parameters are often ignored, especially the
mean radiant temperature. This problem has been discussed by Wray (36) who
defines an "equivalent uniform comfort temperature" as a single thermal
index. This can be used as an indicator of the level of thermal comfort in
the non-uniform conditions of most passive buildings.

Everett (37) draws attention to the very real problem of trying to determine
whether or not a passive system is actually saving any energy. He suggests
a minimum control group of twenty houses and feels that it may be very
difficult to tell if a 'one-off' passive solar house is working or not, or
why it is, given that it does seem to.

Correct control techniques with passive systems could greatly reduce the
initial costs of the building according to a computer analysis carried out
at the University of California (38). Savings of up to 50% in back-up
requirements or, alternatively, the provision of the same amount of stored
solar heat with thinner, smaller walls was reported. With no back-up system
but with a proper control strategy, a thin wall would provide a better
performance than a standard wall of double thickness.

PASSIVE DESIGN CONCEPTS - AN APPRAISAL

With the deeper understanding of passive systems gained with the more
sophisticated monitoring and simulation techniques of the past few years it
has been possible to identify the most desirable characteristics of all the
best known systems. The list chosen by Artese and Barra (39) of the
University of Calabria, Italy, is particularly comprehensive:

   (i)   high solar energy absorption and utilisation efficiency;
   (ii)   high annual contribution from solar heating to total heating
          requirements;
  (iii)   low cost;
   (iv)   low maintenance requirements;
    (v)   high reliability;
   (vi)   long life;
  (vii)   automatic adjustment for different operating conditions, at least
          for daily variations;

(viii) limited use of complicated and fragile devices such as large
        movable summer screens or night insulation;
 (ix)   limited variations in daily internal temperatures;
  (x)   no internal cold or overheated radiating surfaces;
 (xi)   comfortable summer behaviour;
(xii)   easy construction techniques;
(xiii) no significant limitation in north-south building depth, up to a
        reasonable length, e.g. 10-12 m;
(xiv)   suited to normal building methods, both traditional and prefabricated;
        and
(xv)   suited for use in both single and multi-storey buildings.

The results of a series of tests carried out over a period exceeding a year
in a two-storey prototype built at Salisano, Italy, indicated that all these
requirements could be satisfied. It is a natural convection air system with
the following particular features:

(a) a small thermal capacity absorbing surface decoupled from the south
     wall (in the northern hemisphere) of the building. This maximizes
     the heat transfer between the air and the absorber;
(b) the external surfaces of the south wall are insulated in the same way
     as the other walls;
(c) heat distribution to the north side of the house is by air flow
     through channels in the concrete ceiling structure; and
(d) energy storage is in the concrete ceilings. This gives a normal
     building structure a dual function. A feature of the house is that
     the entire building acts as a thermal store because of the external
     insulation.

The very close agreement between the predicted performance and the actual
results, within three percent, confirms Beckman's comments (40) that if the
results from your instruments are more than 5% from the results of a
validated computer model, then the instruments are probably wrong! With
70-75% of the total heating demand met by solar for an additional building
cost of only 7%, the economic advantages of this system are beyond dispute.

It has been shown in this chapter and in Chapter 4 that solar space heating
in buildings can save very considerable amounts of energy. Investment in a
solar energy system is always subject to interference by Government as fuel
prices are raised or lowered, but if it is considered to be socially
desirable to have buildings at least partially heated by solar energy, it is
a function of government to see that it is economically attractive.

REFERENCES

(1) Mazria, E., *The Passive Solar Energy Book*, Rodale Press, Emmaus, 1979.

(2) Stromberg, R. F. and Woodall, S. O., Passive Solar Buildings: A
     Compilation of Data and Results, SAND77-1204, Sandia Laboratories,
     Alberquerque, New Mexico, 1977.

(3) Curtis, E. J. W. and Komedere, M., The heat pump, *Architectural Design*,
     June 1956.

(4) Curtis, E. J. W., Solar energy applications in architecture, Department
     of Environmental Design, Polytechnic of North London, February 1974.

(5) Summers, Royston. Private communication.

(6) Davies, M. G., Model studies of St. George's School, Wallasey, *JIHVE*
39, 77, July 1971.

(7) Davies, M. G., Sturrock, N. S. and Benson, A. C., Some results of
measurements in St. George's School, Wallasey, *JIHVE* 39, 80, July 1971.

(8) Davies, M. G., The contribution of solar gain to space heating, *Sun at
Work in Britain* 3, June 1976 and *Solar Energy* 18, 361-367, 1976.

(9) Climatisation des Habitations Bilan Schématique des Réalisations
1956-1972, CNRS Groupe des Laboratoires d'Odeillo (Pyrénées Orientales).

(10) Robert, J. F., Solar energy work in France, Conference on Solar Energy
Utilization, UK ISES, July 1974.

(11) Trombe, F., Robert, J. F., Cabanat, M. and Sesolis, B., Some
performance characteristics of the CRNS solar houses, ISES Congress,
Los Angeles, Extended Abstracts, Paper 42/15, July 1975.

(12) *Scientific American*, 13 May 1882.

(13) Kelbaugh, Doug, Private communication.

(14) Carriere, D. and Day, F., *Solar Houses for a Cold Climate*. John Wiley,
Canada, 1980.

(15) MacGregor, A. W. K., A solar skin for solid walls, UK ISES Conference
(C19) on The passive collection of solar energy in buildings, April
1979.

(16) MacGregor, A. W. K., The potential for solar space heating in Scotland,
International Journal of Ambient Energy, 1, 149, 1980.

(17) Lawand, T. A., Le Normand, J., Skelton, A. and Papadopoli, N.,
Development of solar wall collectors using forced ventilation, *Solar
World Forum*, Vol. 1, 433-437. Pergamon Press, Oxford, 1982.

(18) Fuchs, R. and McClelland, J. F., Passive solar heating of buildings
using a Transwall structure, *Solar Energy* 23, 123-128, 1979.

(19) McClelland, J. F., Mercer, R. W., Hodges, L., Szydlowski, R. F.,
Sidles, P. H., Struss, R. G., Hull, J. R. and Block, D. A.,
Transwall – A modular visually transmitting thermal storage wall –
status report, *Solar World Forum*, Vol. 3, 1881-1888. Pergamon Press,
Oxford, 1982.

(20) Keable, J., The development of the double envelope concept, *Solar
World Forum*, Vol. 3, 1954-1958. Pergamon Press, Oxford, 1982.

(21) Saito, Y., A simple solar air conditioning system using double roof,
Solar World Forum, Vol. 3, 2160-2164. Pergamon Press, Oxford, 1982.

(22) Hay, H. R. and Yellott, J. I., A naturally air-conditioned building,
Mechanical Engineering 92, 19-25, January 1970.

130   Sun Power

(23) Hay, H. R., Roof, ceiling and thermal ponds, ISES Congress, Los
      Angeles, Extended Abstracts, Paper 41/16, July 1975.

(24) Niles, P. W. B., Thermal evaluation of a house using a movable-
      insulation heating and cooling system, ISES Congress, Los Angeles,
      Extended Abstracts, Paper 42/1, July 1975, and in *Solar Energy* 18,
      413-419, 1976.

(25) Haggard, K. L., The architecture of a passive system of diurnal heating
      and cooling, ISES Congress, Los Angeles, Extended Abstracts, Paper
      47/5, July 1975, and in *Solar Energy* 19, 403-406, 1977.

(26) California Polytechnic, Research evaluation of a system of natural air
      conditioning, HUD contract no. H. 2026 R., January 1975.

(27) Steel, A. F. and Michaelis, D., Passive cooling at the Samuel Jackman
      Prescod Polytechnic, Barbados, (latitude 13°N), *Solar World Forum*,
      Vol. 3, 2170-2173. Pergamon Press, Oxford , 1982.

(28) Greenwood, P. and Ward, H., Solar houses for the elderly, Acorn House,
      Wirral, A development by Merseyside Improved Houses, UK ISES Conference
      (C19), April 1979.

(29) Justin, B., Guy, A. G. and Shaw, G., Monitoring solar walls in occupied
      houses, UK ISES Conference (C24) on modelling and monitoring solar
      thermal systems, London, October 1980.

(30) Justin, B., Alderson, J. V., Guy, A. G. and Shaw, G., Some experience
      gained in the UK from monitoring occupied houses incorporating solar
      walls, *Solar World Forum*, Vol. 3, 1911-1914. Pergamon Press, Oxford,
      1982.

(31) Charfield, J., The Pennyland Demonstration Project, *Solar World Forum*,
      Vol. 3, 1823-1827. Pergamon Press, Oxford, 1982.

(32) Horton, A. R., Linford Passive Solar Houses, *Solar World Forum*, Vol. 3,
      1828-1834. Pergamon Press, Oxford, 1982.

(33) Lee, J. B. G. and Fitzgerald, D., Comparison of physical and
      mathematical models of solar walls, UK ISES Conference (C24) on
      Modelling and monitoring solar thermal systems, London, October 1980.

(34) Besant, R. W., and Schoenau, G. J., Preliminary performance results
      from fourteen commercially designed and contracted low energy passive
      solar houses, *Solar World Forum*, Vol. 3, 1979-1984. Pergamon Press,
      Oxford, 1982.

(35) Kimura, K., Design and operation performance of a hybrid solar house
      without auxiliary space heating, *Solar World Forum*, Vol. 3, 2025-2029.
      Pergamon Press, Oxford, 1982.

(36) Wray, W. O., A simple procedure for assessing thermal comfort in
      passive solar heated buildings, *Solar Energy* 25, 327-333, 1980.

(37) Everett, R., How do we know if we've saved any energy?, *Solar World
      Forum*, Vol. 3, 1792-1796. Pergamon Press, Oxford, 1982.

(38) Sebald, A. V., Clinton, J. R. and Langenbacher, F., Performance effects
      of Trombe wall control strategies, *Solar Energy* 23, 479-487, 1979.

(39) Artese, G. and Barra, O. A., Experimental results and computer models of a new natural convection solar passive system. Proc. Int. Conf. Solar Energy Heating and Cooling, Université Paul Sabatier, Toulouse, 22-24 October, 1981.

(40) Beckman, W. A., Invited lecture. *Ibid.*

CHAPTER 6

THERMAL POWER AND OTHER
THERMAL APPLICATIONS

SOLAR POWERED HEAT ENGINES

The first law of thermodynamics is often expressed as follows:

In any enclosed system, the change of internal energy of the system is equal to the net amount of heat transferred to the system (Q) less the net external work done by the system (W). If E_2 and E_1 represent the initial and final internal energy of the system then

$$Q - W = E_2 - E_1 \qquad (6.1)$$

To obtain a continuous work output it is necessary to bring the system back to its original state, i.e. it must pass through a cycle of operations. In equation (6.1) the net amount of heat transferred, Q, is composed of two parts. Q_1 is the heat supplied at a higher temperature than Q_2, which is the heat rejected at a lower temperature. This is a consequence of the second law of thermodynamics which states:

It is impossible to construct a device which will operate in a cycle and perform work while exchanging energy in the form of heat with a single reservoir.

The higher temperature reservoir is often called a source and the lower temperature reservoir a sink. In another form the second law states that heat transfer can only take place from a hotter to a cooler body.

The efficiency of the cycle is the net work output W divided by the heat input Q_1

$$\eta = \frac{W}{Q_1} . \qquad (6.2)$$

As the operation is cyclic, $W = Q_1 - Q_2$, and the efficiency can also be expressed as

$$\eta = \frac{Q_1 - Q_2}{Q_1} \quad \text{or} \quad 1 - \frac{Q_2}{Q_1} . \qquad (6.3)$$

If the absolute temperature of the source is T_1, and the sink T_2, then the cycle efficiency becomes:

$$\eta = 1 - \frac{T_2}{T_1} \, . \tag{6.4}$$

This is known as the ideal cycle or Carnot efficiency, after the Frenchman Sadi Carnot who was the first to develop these concepts in 1824. A more detailed treatment of this topic and its application to solar power is given by Brinkworth (1).

No engine can have an efficiency greater than the Carnot efficiency. There are various reasons for this, the main ones being losses due to friction between the moving parts and the need for a temperature difference between the source and the engine, and between the engine and the sink, so that heat transfer can take place. In practice it is very valuable to use the Carnot efficiency on a comparative basis, bearing in mind that at best the efficiency of a real engine will be about two-thirds of the Carnot efficiency.

From equation (6.4) it can be seen that the higher the source temperature T_1 becomes, the greater the efficiency for any fixed sink temperature. When this concept is considered with the characteristics of solar collectors, shown in Fig. 3.24, it can be seen that there is a conflict, as any increase in collector temperature results in a corresponding decrease in overall collector efficiency. For any given incident radiation and sink temperature it is possible to construct curves giving the ideal solar engine efficiency, which is the product of the collector overall efficiency, as shown in Fig. 3.24, and the Carnot efficiency. This has been done in Fig. 6.1 for three different collectors, taking a high incident radiation value of 900 W/m^2 and a sink temperature of 300 K (27°C).

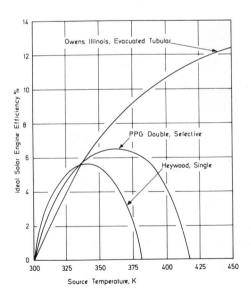

Fig. 6.1.

Up to a 35°C temperature difference between the source and the sink, the
three systems have a very similar performance, indicating that for a highly
inefficient operation at about 2% overall efficiency, a simple collector is
quite adequate. For real overall efficiencies approaching 10% an advanced
collector design is essential or some form of focussing or concentrating
collector system should be provided.

Some Practical Engines

In a summary of work carried out prior to 1960, Jordan (2) commented that
many ingenious solar-powered cyclic devices involving the expansion,
contraction or evaporation of a solid, fluid or gas and capable of
translating this effect into a periodic mechanical motion have been
constructed or proposed. Most of these systems were designed for water
pumping, as there is a tremendous demand for low-cost irrigation in arid
regions, where there is usually a high radiation level throughout the year.

The University of Florida has been a major centre for small scale solar
power generation and their work has concentrated on the development of
fractional horsepower engines (3). Three main types have been studied:

Closed cycle hot air engines in which an enclosed volume of air is displaced
by a piston between hot and cold surfaces. A power piston is operated by
the cyclic high pressures in the cylinder.

Open cycle hot air engines which take in atmospheric air, compress it and
heat it by solar energy. The high temperature and pressure air then expands
and the cycle is completed by an exhaust stroke.

Vapour engines which can operate from flat plate collectors using a
conventional refrigerant, R-11 (trichloromonofluoromethane).

The closed cycle engine had an output of about 240 W with a 1.5 m diameter
parabolic mirror, giving an estimated overall efficiency of just under 20%.
The twin cylinder V-2 solar vapour engine (4) used three 2.8 m^2 flat plate
collectors with an average collecting efficiency of over 50%. The maximum
output was just under 150 W giving an overall efficiency of about 3.5%,
which is in very good agreement with predictions from the previous section.

The University of Florida has also developed a very simple solar pump in
which the only moving parts are two non-return valves (5). A boiler is
connected by a U-tube to a vessel containing non-return valves at inlet and
outlet. The inlet valve section is connected to the water which is to be
pumped. The water in the boiler is heated and turns into steam, forcing
water through the outlet valve from the vessel. When the steam reaches the
bottom of the U-tube, it passes rapidly into the vessel and condenses,
causing the inlet valve to open as a vacuum is formed. This system is a
modern version of Belidor's Solar Pump, illustrated in Fig. 1.1. Another
version has been developed in the UK by AERE Harwell (6) and has a very
simple closed cycle hot air cylinder instead of the boiler used by the
University of Florida. A variant of the basic system, the 'Fluidyne 3'
pump, is shown in Fig. 6.2. One end of the U-tube is heated and the
resulting change in air pressure causes the water in the output column to
oscillate, forcing water through the outlet valve and drawing fresh water
into the system through the inlet valve. As long as the heat is applied,
the pump will continue to oscillate at its resonant natural frequency. A
solar pump developed in India (7) uses pentane vapour generated under

pressure in a flat plate solar collector as the power source. Both an air-
and a water-cooled version are being studied.

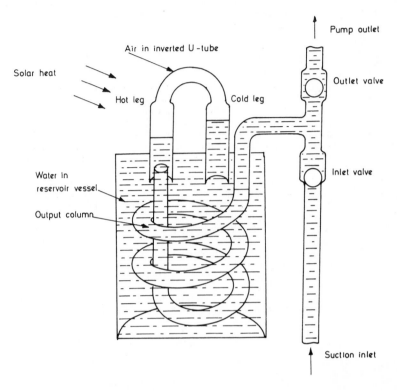

Fig. 6.2. 'Fluidyne 3' pump.

Later studies (8, 9) concentrated on simplifying the system, so that the
vaporized mixture displaces water from one tank to another during the day,
compressing air which is used to pump water from a tank located in a well
through a lift of about 10 m.

Another intermittent system using a diaphragm pump powered by a single
Rankine cycle (10). Solar radiation is collected on flat plate collectors
and vaporizes liquid freon-113. The vapour pushes the diaphragm which, in
turn, pumps water. The exhaust vapour is condensed and then passed into the
flat plate collector by gravity for use during the next day's operation. The
system, which is illustrated in Fig. 6.3 needs some two to three hours
bright sunshine to heat the pump body and generate sufficient vapour pressure.
Overall efficiencies at average solar radiation levels of about 600 W/m^2 are
in the order of 0.3%. The pump is considered to be particularly suitable
for low-lift irrigation where heads of only 3-4 m are required.

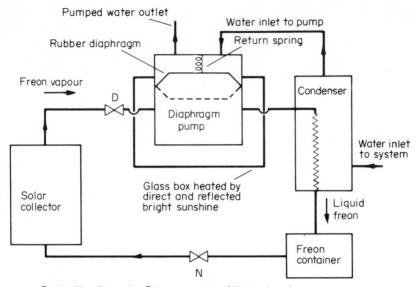

During the day, valve D is open, valve N is shut and
liquid freon collects in the container.
At night, valve D is shut, valve N is open and
allows the liquid freon to charge the solar collector.

Fig. 6.3.   Intermittent Rankine Cycle Solar Pump.

The current state of the art in 1978 was reviewed by Bahadori (11) while
Pytlinski (12) gave an overall historical survey of the previous work. The
economic feasibility was assessed in an analysis carried out in Israel in
1981 (13), when the main conclusion was that as both base energy costs and
the level of inflation have been increasing steadily with the developing
shortage of fossil fuels, solar pumping is becoming an increasingly economic
competitive alternative. An up-dating on the state of the art in 1981 was
given by Derrick *et al.* (14), who have been associated with a UNDP and World
Bank project on small-scale solar powered pumping systems in Mali, the
Philippines and the Sudan.

Although the normal working substances in heat engines are air or steam,
certain metal alloys have the ability, when their shape is changed under
stress, to return to their original shape on being heated. A nickel-
titanium alloy, 'nititol', has this property at about $65^{\circ}C$, a temperature
easily achieved by solar heating. An elegantly simple water pump based on
this concept has been demonstrated in London by Frank and Ashbee (15) as a
result of earlier work on the properties of glass ceramics. The water pump
itself is no more than a ladder of intercommunicating scoops powered by the
basic engine, which is shown in Fig. 6.4. The engine rests on the rim of a
vessel containing water and consists of two vertical pillars each rigidly
mounted to a horizontal axle, with the pillars each linked to a lower
horizontal rigid bar through oppositely flexed leaf springs and oppositely
bent 'nititol' wires. When either of the 'nititol' wires is heated above
$65^{\circ}C$ it tends to straighten, displacing the centre of gravity to the opposite

side of the axle and causing the device to rotate about the axle.   If, as
shown in Fig. 6.4, the axle is supported on the rim of an open vessel filled
with warm water, the device oscillates about the axle, the oscillation being
caused by the 'nititol' wires alternately straightening and displacing the
centre of gravity as they approach or dip into the water.   A somewhat more
elaborate system has been developed in the United States by Banks (16),
while the Copper Development Association in the UK were encouraging
experiments with a copper-based 'memory metal' which became commercially
available in the early 1980s.

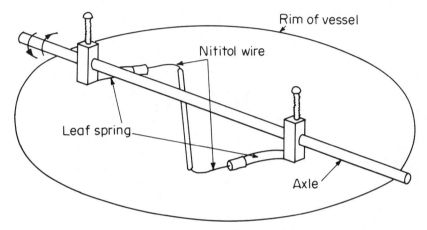

Fig. 6.4.   The Nititol engine.

Furnaces

The most effective type of optical system for a solar furnace is the
parabolic concentrator.   There are severe practical difficulties in making
a large parabolic mirror track the sun, and the alternative method is to
mount the parabolic mirror in a fixed position with its axis horizontal and
facing north (in the northern hemisphere).   Opposite the parabola is a
heliostat which tracks the sun.   This method was used by Trombe (17) at the
Laboratoire de l'Energie Solaire for his first major furnace, which had a
mirror diameter of about 10.7 m, at Mont Louis in the Pyrenees in the
1950s.   The French Centre National de la Recherche Scientifique subsequently
built a 1000 kW furnace at Odeillo (18), probably the best known solar
furnace in existence in the 1970s.   The parabolic mirror, which is 39.6 m
high and 53.3 m wide, contains 9500 individual mirrors with a total reflecting
area of 1920 m^2.   It faces a field of 63 heliostats, with a total mirror
area of 2839 m^2.   Solar furnace research has been reported from the Soviet
Union since the 1960s (19) and by 1981 several countries, including the USA,
Japan and Chile, were also active (20).

Experiments with solar furnaces have proved that it is possible to prepare
refractory oxides at temperatures greater than 3000°C.   Other applications
include the chemical vapour deposition of materials such as molybdenum and
tungsten borides (21) and high temperature phase change studies (22, 23).
The study of the resistance of materials to thermal shock is another
application where the ability of the solar furnace to provide extremely
rapid high temperatures is essential.

In a major review of the use of solar furnaces published in 1981, and citing
78 references, Suresh *et al.* (20) identified many other research areas,
including crystal growth and purification as well as a wide range of
industrial applications.

The philosophy behind all solar furnace research is that the majority of
industrial chemical processes are based on heating by fossil fuels and it
would be valuable to replace these fuels by concentrated solar radiation.
However, although the Odeillo furnace has been shown to be of considerable
importance as a research tool in investigations of the properties of
materials at high temperatures, there are no indications that solar furnaces
will ever be manufactured in quantity. Industrial applications at
relatively low temperatures, such as the baking of bricks, or in replacing
charcoal in pottery kilns, could be of considerable interest to subtropical
countries with low fossil fuel resources.

Solar furnace research has also resulted in the development of two of the
large scale power systems discussed in the following section, the central
'Power Tower' systems and the distributed collector system.

LARGE-SCALE POWER GENERATION

Of the six different approaches to the large-scale generation of solar-thermal
power considered in the following sections, several have attracted
substantial research and development efforts and at least two of the systems
could make a major impact on world energy supplies.

The Central 'Power Tower' System

A central collector system consists of a large field of steered mirrors
which reflect solar radiation to a single central receiver mounted on a
large tower. The solar radiation can be highly concentrated and high
temperature steam can be generated in the receiver. The possibility of
using alternative heat transfer fluids is also being considered. A 50 kW
pilot plant was built during the 1960s at St Illaria-Nervi in Italy and can
produce 15 kg/h of superheated steam at 500°C, with a collector array
consisting of 270 mirrors each one m in diameter (24).

During the mid 1970s studies in the USA (25-27) considered individual units
of 100 MW capacity for intermediate and peaking loads with the height of
the tower ranging from 300 to 450 m. Three test facilities had been
developed, a 30 kW solar furnace in White Sands, New Mexico; a 400 kW unit
at the Georgia Institute of Technology and the Sandia Laboratories 5 MW
Centre Receiver Test Facility (CRTF) in Albuquerque, New Mexico (28).

The next step towards the 100 MW concept was the commissioning of a 10 MW
(electrical) plant at Barstow, California, in January 1982. In Europe, the
EEC and the IEA (International Energy Agency) have complementary programmes,
all scheduled for completion by the end of 1982 as indicated in Table 6.1.

Table 6.1.

Organization and date of planned completion	Location	Electrical output and type	Collector details	Output temperature
EEC 1981	Sicily	1 MW, Central Tower	7800 m^2 Two types of heliostat	450°C Water/ steam
IEA 1982	Almeria, Spain	500 kW, Central Tower	6150 m^2 Three types of heliostat	530°C Sodium
IEA 1982	Almeria, Spain	500 kW, Distributed	13 000 m^2 Three types of collector	295°C Sodium

The aim of these programmes was to examine as many systems as possible before proceeding to fully commercial systems. Other countries with separate MW programmes for the early 1980s included Spain (1 MW at Almeria), Japan (two 1 MW plants at Shikoyu) (29), France (1.5 MW at Targansonne), and a further 25 MW unit, the gas-cooled plant designed by Interatom (Germany) for Spain, with a proposed tower height of 200 m and a mirror area of 120 000 m^2. The 1 MW plant in Sicily is shown in Fig. 6.5.

The Distributed Collector System

Large numbers of individual collectors are the feature of this system, which has been called a 'solar farm'. An extensive system of insulated pipework is necessary to collect the energy centrally to the generating station. This could be an appropriate application for the SRTA collector, discussed below. An impression of the distributed collector system is illustrated in Fig. 6.6. As an alternative to individual collectors, long parabolic troughs could be used. Both systems could be located in desert areas, but their use is limited to regions with large amounts of direct radiation.

The Stationary Reflecting/Tracking Absorber (SRTA)

The collector, which is shown in Fig. 6.7, consists of a segment of a spherical mirror placed in a stationary position facing the sun. It has a linear absorber which can track the image of the sun by a simple pivoting motion about the centre of curvature of the reflector (30, 31). From experience gained with pre-production prototypes it has been estimated that for large-scale applications in the range from 10 to 100 MW it would be feasible to produce electric solar power at less than the projected costs for nuclear power. Among the advantages mentioned for the SRTA in building applications are that the same system can be used for electric power and hot water, the working fluid can reach a high temperature, thus reducing the size of the storage system, and that there is no danger of breakages with large glazed areas. The main disadvantage in countries with a high proportion of diffuse radiation is that it is designed to absorb direct radiation and will not absorb much diffuse radiation. It has very considerable potential in countries with a high proportion of direct radiation for various small-scale power applications.

Fig. 6.5.   EURELIOS – the 1MWe plant in Sicily.

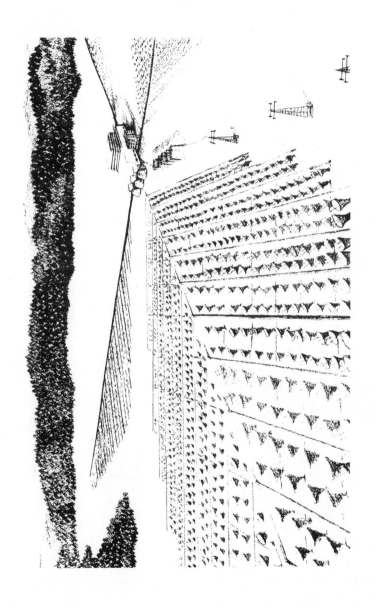

Fig. 6.6.

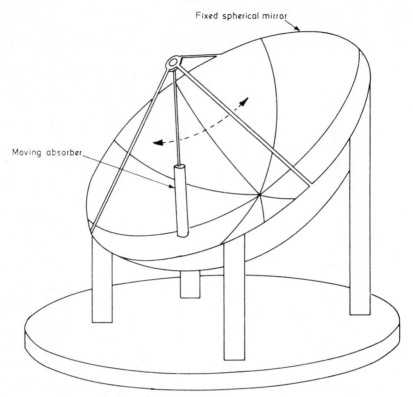

Fig. 6.7.   The Stationary Reflecting/Tracking Absorber
            (SRTA).

Ocean Thermal Energy

The use of the temperature difference between the surface of the ocean and
the colder deep water to operate a heat engine was proposed towards the end
of the nineteenth century. The oceans are natural solar energy collectors
and require no special storage systems or manufactured collectors and,
because of their enormous size, have considerable potential to compete
economically against other methods of power generation. The earliest
system was a 22 kW power plant off the coast of Cuba developed by Claude
(32) in the late 1920s. It had an overall efficiency of less than 1% and
operated with an open Rankine cycle in which the higher temperature sea
water was passed directly to a low pressure evaporator to provide steam to
power the turbine. It was uneconomic at that time, as was a subsequent
larger project by the French some twenty years later, and further work was
discontinued.

Renewed interest in the concept came in the 1960s when the possibility of
using a closed Rankine cycle was suggested (33) in the USA. This work formed
the basis for several major large-scale theoretical investigations which
were summarized by McGowan (34) in 1976. Five different research teams from
industrial organizations and universities selected different systems for
their various site locations. The net power outputs ranged from 100 to
400 MW, with overall ocean temperature differences of 17.8°C for the Gulf

Stream off Miami to 22.2°C for a tropical location (within 20° of the equator). Three different working fluids were suggested, R-12/41, propane and ammonia. The overall net cycle efficiencies of all five proposals were very similar, ranging from 2.1 to 2.4%. Environmental studies have also been carried out, but these have been concerned with the effects of the environment on the power plant rather than the effect of the plant on the surrounding environment. Various studies of possible ocean sites were also carried out at this time. For example, a report on the thermal properties of the Florida Current suggested that the total available ocean thermal energy delivered in the form of electricity per annum could be 315×10^{12} kWh (35).

Following these systems feasibility studies, a joint project called mini-OTEC was developed by the State of Hawaii, Lockheed Missiles and Space Company and the Dillingham Corporation. By 1979 a 50 kW OTEC plant was operating off the coast of the island of Hawaii. Its performance closely followed the theoretical prediction (36). The two key reference sources to the various development programmes have been prepared by Avery and Dugger of the Ocean Energy Programme at the Johns Hopkins University (37, 38). As early as 1976 they were predicting that the first commercial plants producing ammonia for fertilizers could be deployed by the mid-1980s, while direct economically competitive delivery of electric power could follow by 1990. OTEC plant-ships of 325 MW are envisaged.

By the start of the 1980s, several other countries apart from the USA had active OTEC programmes. These included France, Japan and Eurocean, a European consortium (39). A fascinating modification of OTEC technology was suggested by Professor Estefan of the National Research Centre, Cairo (40). He pointed out that the Red Sea contains some zones of the hottest and saltiest water in the world within its volume of approximately 21 500 km^3. Below a transition zone between 100 and 400 m deep and above the deeper hot brines, the water temperature is reasonably uniform at 22°C. The intermediate brines and deep brines at levels of some 2000 m have temperatures ranging from about 44 to 56.5°C. Professor Estefan suggested that it appeared to be technically feasible to use an 'inverted' OTEC technique to utilize this temperature gradient.

Satellite Solar Power Station

The use of a satellite in orbit round the earth to produce electricity which could be fed to microwave generators for transmission to earth was first proposed by Glaser in 1968 (41). Since then it has attracted several detailed feasibility studies in the USA. The basic principle is photovoltaic solar energy conversion, after concentration, from two symmetrically arranged solar cell arrays. The microwave generators form an antenna between the two arrays, which directs the microwave beam to a receiving antenna on earth. In synchronous orbit the satellite would be stationary with respect to any particular location on earth and full use could be made of practically continuous solar radiation. Up to 15 times more energy is potentially available on this basis, compared with terrestrial applications which are limited by weather conditions and daily cycles. The system could be designed to generate from 3 to 15 GW (42).

A full assessment of the concept has been made in the USA by the Department
of Energy and NASA and, on a smaller scale, in Europe by the European Space
Agency (43, 44). By the early 1980s no single constraint had been identified
which would preclude continuation of research and development for either
technical, economic, environmental or societal reasons.

Heliohydroelectric Power Generation

The concept of heliohydroelectric power generation is to convert solar energy
into electricity by first transforming it into hydraulic energy. If a closed
reservoir is completely sealed off from the sea, the level of the reservoir
will tend to decrease as a result of evaporation. Hydroelectric generators
could be placed at the reservoir end of pipes connecting the reservoir to
the sea. The fall in the level of the reservoir induces a flow of water
from the sea, and the potential energy caused by the difference in water
levels could be transformed into electrical power. By choosing suitable
water levels and power systems, it would be possible to have a continuous
process. This topic has been extensively studied in Saudi Arabia by Kettani
(45, 46), who has measured evaporation rates and compared these with
meteorological data. The possibility of building a dam across the Gulf of
Bahrain is being explored (47), using the whole sealed-off Bay to create a
hydraulic head from the open sea in the Gulf.

Connecting the Mediterranean Sea with the Qattara Depression for a
hydroelectric scheme was first considered in 1916. This project was revived
during the 1970s by a joint Egyptian-Federal Republic of Germany feasibility
study. The results of this study, which was completed in December 1980,
have been reported (48) but a firm decision on whether or not to proceed
with the scheme was not anticipated until 1982 at the earliest.

OTHER FOCUSING SYSTEMS

Cookers

Cooking by solar energy has attracted several groups of research workers
since successful cookers were reported in the eighteenth and nineteenth
centuries. Solar cookers can be classified into three groups. The earliest
was the 'hot-box' or simple solar oven which consisted of an insulated box
with a matt black interior covered with at least one transparent cover plate.
Later versions had hinged lids with reflecting surfaces. In good radiation
conditions temperatures greater than 100°C can be held for several hours.
The second group uses some type of focusing system to concentrate the
radiation. In the 1920s Abbot (49) used a combination of a parabolic mirror
and oil as the heat transfer fluid so that cooking could be achieved well
into the evenings, because the higher temperatures reached by the oil gave
an increased storage capacity. An extensive series of tests in India at the
National Physical Laboratory under Ghai (50) led to the design of a
reflector-type direct solar cooker with a spun aluminium parabolic reflector.
Full details of the manufacturing process were also given (51). Aluminized
plastic film was successfully used in several cookers developed at the
University of Wisconsin (52), one of which was a portable folding type
mounted on a standard umbrella frame. Spherical or cylindrical concentrators
have also been used and developments of the hot box and concentrating types
have been reported at the University of Florida (5).

The third group uses the flat plate collector as a solar steam cooker and consists of two main components, the collector and an insulated cooking chamber, which is essentially a steam bath in which the cooking vessel can be placed. The collector has two or three transparent cover plates and has vertically rising pipes bonded to a metal sheet. These pipes are connected directly to the cooking chamber at the top of the collector. A collector with overall dimensions of 0.8 m wide by 1.55 m long was connected to a chamber containing a single 200 mm diameter by 125 mm deep aluminium cooking vessel in Haiti (53). This collector was based on a smaller unit developed at the Brace Research Institute (54). Spun aluminium parabolic cookers are commercially available and Fig. 6.8 shows one displayed at the ISES World Congress in Los Angeles.

Fig. 6.8.

An innovative development was described in 1981 by Walton (55) of the Georgia Institute of Technology, Atlanta. A lightweight spiral-shaped point focusing collector was formed by cutting a spiral pattern from a sheet of flat material. This material is then attached to a flat background sheet. By using simple materials such as hardboard and aluminium foil, a low cost (less than $5) concentrator with a diameter of 1.1 m was constructed. Tests were planned for Mali. The importance of an interdisciplinary approach, including familiarity with the socio-cultural norms affecting cooking has been stressed (56). Future development of cookers may use the heat pipe to transfer the heat from the collector to a longer term storage unit, so that cooking can be carried out in the late evening or early morning, but the importance of low-cost units cannot be overemphasized.

The Fresnel Lens

The concentration ratio which can be achieved by a single lens is limited, as
it is difficult to manufacture accurate conventional lenses with very short
focal lengths and the concentration ratio is proportional to the ratio of the
diameter of the lens to its focal length. The Fresnel lens combines the
advantages of a multi-lens system within a single unit as each segment is
designed to concentrate the incident radiation onto a centrally positioned
receiver. A further advantage is the Fresnel lens is relatively thin in a
direction normal to the radiation. Figure 6.9 shows the cross-section of a
linear Fresnel lens, which can be mounted in a large array with one
dimensional tracking (57). The performance characteristics of several linear
Fresnel lens systems has been reported (58, 59) and they appear to be
superior to the evacuated tubular type for direct radiation up to temperatures
of about 250°C. The long term predictions for the cost of Fresnel lens
systems show that they would be very competitive (58). A circular Fresnel
lens has been developed for low concentration in photovoltaic cells (60).

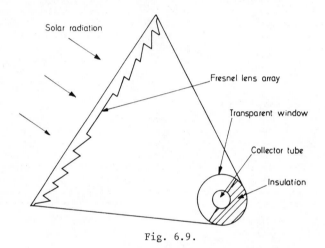

Fig. 6.9.

REFRIGERATION AND COOLING

The great advantage of using solar energy in refrigeration and cooling
applications is that the maximum amount of solar energy is available at the
point of maximum demand. There are two quite different major applications,
the first in the cooling of buildings and the second in refrigeration for
food preservation or for storing vaccines for medical purposes.

In building applications part of the solar cooling system could be used to
provide heating outside the hot mid-summer period and the cost of the system
could be shared between the two functions. In a theoretical analysis of the
more complex systems involving combined heating and cooling for eight cities
in the USA, Löf and Tybout (61) showed that the combined system was more
economical in six of the eight cases studied. The cooling demand is at a
maximum during the early afternoon, depending on the orientation of the
building and its thermal mass, so that the storage capacity for cooling is
only a few hours in contrast to the very much greater periods required for
heating systems.

In solar heating applications the hot fluid from the collectors can often
be used to heat the interior of the building directly, but most solar cooling
applications involve a solar powered engine system. Four main methods have
been adopted as follows:

(i)   the compression refrigeration cycle in which the refrigeration side
        is driven by a solar powered engine;
(ii)   absorption systems;
(iii)   evaporative cooling;
(iv)   radiative cooling.

The basic compression refrigerator is a familiar domestic appliance and the
electric motor would be replaced by a solar powered engine for a
straightforward application of the first method. Several complex compression
refrigerator systems have been tried or suggested, including a proposal (62)
that a four-cylinder reciprocating engine should have two solar-powered
cylinders operating on R-114 driving two refrigeration cylinders using
another fluid, R-22. The Mobile Solar Research Laboratory, a joint project
of the NSF and Honeywell, was fitted with a conventional vapour compression
cooling system operating on R-12, which was driven by a high-speed solar
powered turbine operating on R-113. Test results from this system gave an
overall coefficient of performance of 0.5 with a turbine inlet temperature
of just over 100°C (63). An alternative approach that was commercially
available by the early 1980s was to replace the a.c. electric motor in the
conventional compression refrigerator by a d.c. motor which was powered by
photovoltaic cells.

The basic principles of a closed cycle absorption system for cooling are
shown in Fig. 6.10. The working fluid is a solution of refrigerant and
absorbent. When solar heating is supplied to the generator some refrigerant
is vaporized and a weak mixture is left behind. The vapour is then condensed
and expands to the lower pressure evaporator, where it is vaporized and
refrigerates the external working fluid, which would be air for an air
conditioning application. The cycle is completed in the absorber when the
refrigerant recombines with the original solution and is pumped back to the
generator.

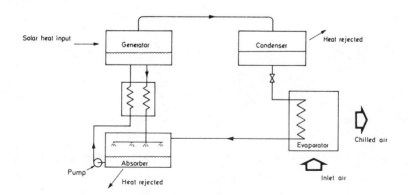

Fig. 6.10.

Various other developments in absorption systems were reviewed during the
mid- to late-1970s (64-67), when solar powered litium-bromide systems became
commercially available.

Among several examples reported in the literature was a space cooling water-
lithium bromide system on the Yazaki Solar House One in Japan (68) which had
a coefficient of performance on a clear day of approximately 0.5. Although
ammonia-water systems had been successfully developed and tested in the
laboratories of the University of Florida (5, 69) and the University of the
West Indies (70) some years earlier, by the end of the 1970s several studies
tended to confirm that water-lithium-bromide systems would be simpler to
operate under identical conditions (67, 71). This view was supported by
experimental studies at the Brighton Polytechnic, UK (72) which revealed
severe practical limitations with ammonia-water systems. The original
advantage of the continuous ammonia-water system was that it could be
operated by the flat plate collectors available at the time, but as more
sophisticated collectors became available, this advantage was no longer
significant. By 1979 Ward et al. (73) at Colorado State University, were
able to report collector efficiencies exceeding 50% under conditions of
$\Delta T/H_T = 75^{\circ}$C m^2/kW, where ΔT = collector fluid inlet temperature minus
ambient temperature and H_T = incident solar radiation on a tilted surface,
with evacuated tubular collectors integrated with two lithium bromide
cooling units installed on their Solar Houses I and III.

Intermittent ammonia-water absorption units are relatively simple to operate
and have been successfully developed at the Asian Institute of Technology
(AIT), Bangkok, by Professor R. H. B. Exell (74). During the day, heat from
solar collectors vaporizes the ammonia and it condenses in the condenser.
At night the ammonia is passed through an expansion valve into an evaporator,
where the refrigeration effect produces ice. By 1981 the principal unit at
the AIT could produce up to 30 kg of ice from water at 28°C, with a
collector area of 5 m^2 linked to the ammonia-water system. The solar
radiation was enhanced by the use of plane reflecting mirrors. A similar
system was reported from the University of Papua New Guinea (75).

A third type, the open cycle, has been analysed in the USA (76), and is
reported to have attracted considerable interest in the USSR. The open
cycle regenerates the weak absorbent solution by losing refrigerant to the
atmosphere instead of to the condenser as in a closed system.

Evaporative systems achieve a cooling effect through the evaporation of water.
A simple method used by Thomason (77) takes water from the house storage
tank and trickles it down the unglazed north roof. Good results have been
achieved in Australia, where water has been evaporated in the air discharged
from the buildings, and this exhaust air chills rocks in a switched-bed
rock-filled recuperator. The air flow is switched every ten minutes and the
incoming fresh air is cooled by the rocks before passing to the building
(78, 79). A completely different approach is the use of a desiccant wheel
system with a solar heat input to supplement a conventional gas fired burner
(80). The first installation, near Los Angeles, was completed in 1975.

Radiative cooling is suitable at night with clear atmospheric conditions and
Yanigimachi (81) and Bliss (82) have used this technique by pumping water
into roof-mounted collectors. This technique has been analysed by Hay (83).
The Institute of Experimental Physics at the University of Naples has
demonstrated that radiative cooling can also take place during the day in
the presence of diffuse solar radiation (84). Selective surfaces have been
prepared with optical properties matched to the radiation emitted by the

atmosphere. This radiation has a minimum value between 8 and 13 µm, forming an 'atmospheric window'. Theoretical predictions indicated that temperatures about 10 to 15°C less than ambient could be achieved. The actual values reached in a small model test were less than this, but confirmed the trend indicated theoretically.

Solar Thermochemical Heat Pumps (85-87)

Solar thermochemical heat pumps are broadly similar to solar absorption refrigerators in that they use high grade heat and reversible chemical reactions to raise low grade thermal energy to some intermediate level. They differ mainly in that they always operate on an intermittent cycle and are capable of almost indefinite storage periods. This makes them suitable for a wide range of applications. The basic theory, using a Carnot analysis, was described in 1978 (85). One possibility which was also explored in 1978 was the use of metal hydrides as chemical heat pumps (86). A major review of the field was carried out by Swet in 1981 (87) who identified representative matchings of various chemical combination, e.g. calcium chloride-methanol with space heating and cooling; sulphuric acid-water with industrial process heat; calcium oxide-water with irrigation pumping and paired ammoniated salts with cooking. Investigations of this latter possibility at the Brighton Polytechnic, UK, showed that the high cost of the container and valves would be a difficult problem to overcome.

SOLAR PONDS

A few days before the end of the most eventful decade in the history of solar energy applications, the world's first solar pond thermal power station, with an output of 150 kW, was switched on. Lights shone out over the Dead Sea at midnight on 16 December 1979. This was the culmination of a quarter of a century's work by Harry Tabor of the Scientific Research Foundation in Israel, who had been the driving force behind the project, and whose Review Paper, published in 1981 (88), gives an authoritative record of solar pond research and development. The 45 references cited in this paper include ten to his own work, the earliest dating from 1961 (89).

The term solar pond in this context refers to a salt-gradient, non-convecting pond, which is a body of saline water in which the concentration increases with depth. In a natural pond when solar radiation heats the water below the surface the action of convection currents causes the heated water to rise to the surface and the pond temperature normally follows the mean temperature of the surroundings. A solar pond contains concentrations of dissolved salts which gradually increase with depth, causing the density of the water to increase towards the base of the pond, which is often black. Solar radiation penetrates to the base, heating the water at this lower level, but any convection currents are suppressed by the density gradient. Heat losses from the surface are reduced, compared with a natural pond, and the temperature at the bottom of the pond rises. While there are daily fluctuations in both ambient air temperature and in the upper water layers, the temperature at the bottom of the pond, where heat would be extracted, remains fairly uniform (90). A solar pond is both a massive heat collector and heat storage system and, compared with a conventional collector and heat store, is relatively inexpensive. Other advantages are:

  (i)   problems which could occur with dirt settling on the surface and
        reducing collection efficiency are eliminated;
 (ii)   extracting energy from the pond is very straightforward as the lowest
        hot layer can be pumped to the power station and returned to the pond
        for reheating;
(iii)   solar ponds can operate continuously throughout the year provided the
        storage capacity is chosen to match the demand;
 (iv)   a solar pond power system can - like a hydroelectric plant - provide
        peaks of power on demand.

The system is illustrated in Fig. 6.11.

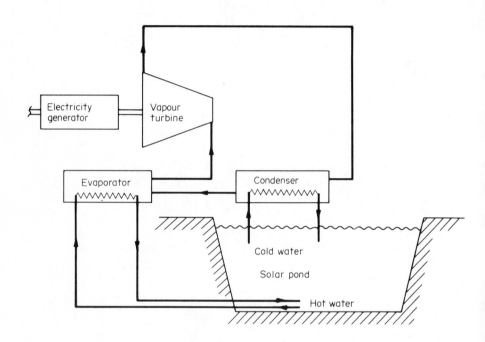

Fig. 6.11.   Solar pond.

The analysis of its performance is complex.   General equations were
determined in the 1960s (91) and several modifications have been reported
since then, including an analysis for a site in the UK (92).   Among earlier
ideas, which are now being re-examined, was a proposal to 'float' a solar
pond in a large saline lake, such as the Dead Sea (93).

Methods of facilitating heat storage and extraction under investigation include the use of transparent membranes submerged beneath the surface of the pond to establish a non-convective salt water layer at the top over a separated convection zone. The concept of a shallow solar pond, about 50 mm deep, with a transparent plastic cover, which could be emptied at night into an underground covered storage reservoir, has also been studied for the production of electricity at the megawatt level (94).

The range of possible applications continues to increase. Space heating for buildings and greenhouses (95, 96), process heating (97), space cooling, desalination and salt production (88) being the more obvious. For power production the next stage in the Israeli programme will be the commissioning of a 5 MW unit by 1983. At least ten other countries have had active research programmes, including India, the USA and the USSR. The economics are potentially very attractive. With the technology for converting low temperature heat into electricity now well established, the unit cost for total power systems in the 20-40 MW range was estimated to be \$3000/kW (1980 prices) installed in 1981, with a possible reduction to \$2000/kW installed for even larger units (88).

DISTILLATION

One of the major problems in many parts of the world is the scarcity of fresh water and the development of inexpensive, large solar distillation units capable of easy transportation and handling is increasingly important. Solar distillation is another application which dates back to the nineteenth century and the simplest form of water still now in use is basically unchanged from the early designs which consisted of a shallow tray, filled with salt or brackish water, and covered by a sloping glass cover plate. The solar radiation heats the water in the tray and evaporates it. When the vapour comes in contact with the colder surface of the glass it condenses, forming fresh water which runs down the inner surface in the form of droplets and can be collected in a trough at the lower edge. Under good radiation conditions an output of about 4 kg/m^2 of fresh water can be obtained daily. Two excellent summaries of solar distillation techniques were published in 1970, the first giving a comprehensive review of the history, theory, applications and economics (98) and the second concentrating on potential applications in developing countries (99).

One country where considerable practical experience has been gained over many years is Australia. A guide to the design, construction and installation of a solar still developed by the Commonwealth Scientific and Industrial Research Organization was published in 1965 (100) and subsequent developments were reviewed nine years later by Cooper and Read (101). A large installation, with an evaporating surface area of 8667 m^2 was completed on the island of Patmos in the Aegean in 1967 (102). The average distillation rate was 3.0 kg/m^2 per day, with a maximum of 6.2 kg/m^2 at mid-summer. The first large installation designed and manufactured in the UK was a 185 m^2 unit for Aldabra in the Indian Ocean (103) in 1970. At least one small company in the USA has been making solar stills for domestic use since the 1950s (104) and among a considerable volume of university research, the University of California, Berkeley, has a record of over 20 years activity. This work is reviewed by Howe and Tleimat (105), who include a design for a 37.85 m^3/day distillation plant, with steam generated at 65.5^{o}C. Fundamental studies were also carried out on the performance of single sloped and double sloped stills under Indian arid zone conditions (106).

Methods of improving the simple design outlined above were described by
Moustafa *et al*. (107) in 1979. They tested three types of still. The first
was a basin type with a series of stepped salt water shelves, a charging
pump and a collection tray for the distilled water. The second type was
similar to the first type, but with the addition of a condenser reservoir
on the shaded side. The third type was a wick collector-evaporator system
with a synthetic wettable mat, one inch (25.4 mm) thick, lined with black
plastic, down which the brine trickled. The wick type proved to have a much
higher overall efficiency than the other two, producing over twice as much
distillate in tests under identical conditions. The superiority of the wick
type was also reported from Japan in 1981 (108), when an increase in
productivity of between 20 and 50% compared with the simple basin type was
confirmed.

Improved performance with simple stills was also achieved by lining the
basin with charcoal pieces (109) to give a gain of up to 15%, while the
addition of water soluble dyes (110) was reported to give gains of up to 29%,
with black naphylamine dye giving the best results. A reverse osmosis water
desalination unit powered by an 8 kW (peak) photovoltaic array provided
drinking water for a community of 250 people near Jeddah, Saudi Arabia, and
was being used in 1981 to check the validity of computer models of the
system (111). Solar powered reverse osmosis and electrodialysis were also
being studied in Saudi Arabia at the King Abdulaziz University (112).

INDUSTRIAL APPLICATIONS

Process Heat

Any increase in the output temperature from a flat plate collector reduces
its overall efficiency and the cost effectiveness of any solar heat
generating system. The potential for the large-scale use of solar energy
for industrial processes is very dependent on the operating temperature of
the process. In Australia, where there is a long tradition of solar
applications and research, a feasibility study of a typical food processing
plant in Melbourne (113) showed that it was technically practicable to phase
solar energy heating systems into existing processes. Over 50% of the
annual heat requirements could be provided by solar collectors using known
techniques, as over 70% of the requirement was at temperatures less than
100°C and there was no significant usage above 150°C. In the food processing
industry, heat is typically generated in a central boiler house at a
temperature higher than that required for any of the processes in the plant,
and then distributed as water at 99°C or low pressure steam, 125-170°C, to
the individual processes, most of which operate at much lower temperatures.
The successful integration of a solar installation with such a system
requires the solar collectors to operate at the lowest practicable
temperature and the collectors, with their associated storage if required,
are coupled directly to individual processes.

Industrial solar heat generating systems must ensure that heat is available
for the manufacturing process at all times and must incorporate sufficient
thermal storage and collector capacity to guarantee this under the worst
operating conditions. Alternatively, auxiliary heating must be available
during periods of low radiation.

Towards the end of the 1970s commercial experience of solar systems matured
and examples of large-scale applications could be found in many countries.
In addition to food processing, service industries such as hotels, laundries
and hairdressing salons began to demand solar heating as it became
competitive with traditional fuels.

Crop drying, a traditional solar use, became more sophisticated with examples such as an air collector system for industrial crop dehydration in California (114) and the drying of wet grains after harvest, using peanuts as the typical crop (115), while low-cost inflated polythene tubes have been successfully used for fruit dehydration (116). A feature of some of the work being carried out in the early 1980s was the response which some institutions were making to the Brandt Report (117). An excellent example of this north-south cooperation was the small scale solar crop drier work carried out by Twidell *et al.* at the University of Strathclyde, UK, and Kenyatta University College, Nairobi, (118).

Among the many different industrial processes which were being examined at the start of the 1980s was the feasibility of solar coal gasification (119). The analysis indicated that coal could be gasified by reacting it with steam (or carbon dioxide) in the focal zone of a solar central receiver plant. The process would have the attraction of upgrading coal to a more easily utilized form of fuel and of chemically storing solar energy. Other projects in the United States Department of Energy programme included enhanced oil recovery, oil refining, fertiliser production and ore treatment.

Transport

The solar-electric car of the University of Florida (5) was the first solar powered car to operate under normal traffic conditions. The car was powered by a 27 h.p. electric motor, which was battery-driven from NiCd and Pb acid batteries. The batteries could be charged either by photovoltaic cells or from a solar engine-generator system. The car had a top speed of 29 m/s on a level road and a range of over 160 km. The long-term proposals are for the establishment of solar battery charging stations, where a discharged battery could be exchanged for a fresh one, providing an energy-free and non-polluting transportation system.

REFERENCES

(1) Brinkworth, B. J., *Solar Energy for Man*, 115-146. Compton Press, Salisbury, 1972.

(2) Zarem, A. M. and Erway, D. D. (ed.), *Introduction to the Utilisation of Solar Energy*, 145. McGraw-Hill, 1963.

(3) Farber, E. A., The University of Florida solar energy laboratory, Conf. The Sun in the Service of Mankind, UNESCO, Paris, 1973.

(4) Farber, E. A. and Prescott, F. L., A solar powered V-2 vapor engine, *Ibid*.

(5) Farber, E. A., Solar energy conversion research and development at the University of Florida, Building Systems Design, February/March 1974.

(6) West, C. D., The Fluidyne heat engine, Proc. Conf. Solar Energy Utilisation, UK Section, ISES, 54-59, July 1974.

(7) Rao, D. P. and Rao, K. S., Solar water pump for lift irrigation, ISES Congress, Los Angeles. Extended Abstracts. Paper 13/12, July 1975, and in *Solar Energy* 18, 405-411, 1976.

(8)   Sudholner, K., Krishna, M. M., Rao, D. P. and Soin, R. S., Analysis
      and simulation of a solar water pump for lift irrigation, *Solar Energy*,
      24, 71-82, 1980.

(9)   Kwant, K. W., Rao, D. P. and Srivastava, A. K., Experimental studies
      of a solar water pump, *Solar World Forum*, 2, 1172-1176. Pergamon
      Press, Oxford, 1982.

(10) Sharma, M. P. and Singh, G., A low lift solar water pump, *Solar Energy*
      25, 273-278, 1980.

(11) Bahadori, M. N., Solar water pumping, *Solar Energy* 21, 307-316, 1978.

(12) Pytlinski, J. T., Solar energy installations for pumping irrigation
      water, *Solar Energy* 21, 255-262, 1978.

(13) Karmeli, D., Atkinson, J. F. and Todes, M., Economic feasibility of
      solar pumping, *Solar Energy* 27, 251-260, 1981.

(14) Derrick, A., Fraenkel, P. L., McNelis, B. and Starr, M. R., Small scale
      water pumps - the state of the art, *Solar World Forum*,
      1146-1150. Pergamon Press, Oxford, 1982.

(15) Frank, F. C. and Ashbee, K. H. G., Heat engine uses metal working
      substance, *Spectrum* 132, 2-4, 1975.

(16) Banks, R., Nititol heat engines, ISES Congress, Los Angeles, Extended
      Abstracts, Paper 53/4, July 1975.

(17) Trombe, F., Solar furnaces and their applications, *Solar Energy* 1, 9,
      1957.

(18) Trombe, F. and Le Chat Vinh, A., Thousand kW solar furnace built by
      the National Centre of Scientific Research in Odeillo (France), *Solar
      Energy* 15, 57-62, 1973.

(19) Arifov, U. A., Development of solar engineering in the USSR,
      Gelioteckhnika 8, 3, 1972.

(20) Suresh, D., Rohatgi, P. K. and Coutures, J. P., Use of solar furnaces
      - 1, materials research, *Solar Energy* 26, 377-390, 1981.

(21) Trombe, F., Gion, L., Royere, C. and Robert, J. F., First results
      obtained with the 1000 kW solar furnace, *Solar Energy* 15, 63-66, 1973.

(22) Nogucki, T., Mizuno, M. and Yamada, T., High temperature solar furnace
      studies, ISES Congress, Los Angeles, Extended Abstracts, Paper 23/1,
      July 1975.

(23) Mizuno, M., High temperature phase studies on the system $Al_2O_3-Ln_2O_3$
      with a solar furnace, Paper 23/2, *Ibid*.

(24) Francia, G., Pilot plants of solar steam generation stations, *Solar
      Energy* 12, 51-64, 1968.

(25) Sobin, A., Wagner, W. and Easton, C. R., Central collector solar
      energy receivers, *Solar Energy* 18, 21-30, 1976.

(26) Vant-Hull, L. L. and Hildebrandt, A. F., Solar thermal power system based on optical transmission, *Solar Energy* 18, 31-39, 1976.

(27) Blake, F. A., 100 MWe solar power plant design configuration and performance, NSF-RANN Grant No. AER-74-07570, Martin Marietta Aerospace, Denver, 1975.

(28) Moeller, C. E., Brumleve, T. D., Grosskreutz, C. and Seamons, L. O., Central receiver test facility, Albuquerque, New Mexico, *Solar Energy* 25, 291-302, 1980.

(29) Tanaka, T., Solar thermal electric power systems in Japan, *Solar Energy* 25, 97-104, 1980.

(30) Steward, W. G., A concentrating solar energy system employing a stationary spherical mirror and movable collector, Proc. Solar Heating and Cooling for Buildings Workshop, 21-23 March 1973, NSF-RANN-73-004, July 1973.

(31) Kreider, J. F., The Stationary Reflector/Tracking Absorber Solar Collector, Presented at US Section ISES Meeting, Ft. Collins, Colorado, 19-23 August 1974.

(32) Claude, G., Power from the tropical sea, *Mechanical Engineering* 52, 1039-1044, December 1930.

(33) Anderson, J. H. and Anderson, J. H. Jnr., Large-scale sea thermal power, ASME Paper No. 65-WA/SOL-6, December 1965.

(34) McGowan, J. G., Ocean thermal energy conversion - a significant solar resource, *Solar Energy* 18, 81-92, 1976.

(35) Mangarella, P. A. and Heronemus, W. E., Thermal properties of the Florida current as related to Ocean thermal energy conversion (OTEC), *Solar Energy* 22, 527-533, 1979.

(36) White, H. J., Mini-OTEC, *International Journal of Ambient Energy*, 1, 75-88, 1980.

(37) Dugger, G. L., Francis, E. J. and Avery, W. H., Technical and economic feasibility of ocean thermal energy conversion, *Solar Energy* 20, 259-274, 1978.

(38) Avery, W. H. and Dugger, G. L., Contribution of ocean thermal and energy conversion to world energy needs, *International Journal of Ambient Energy*, 1, 1980.

(39) Expanded Abstracts of the 7th Ocean Energy Conference, Washington DC, June 1980.

(40) Estefan, S. F., Modified OTEC technology for the Red Sea area, Paper E6-8:91, ISES Solar World Forum Abstracts. Pergamon Press, Oxford, 1981.

(41) Glaser, P. E., Power from the sun : its future, *Science* 162, 857-861, November 1968.

(42) Glaser, P. E., The case for solar energy, Conf. Energy and Humanity, Queen Mary College, London, September 1972.

(43) Glaser, P. E., Progress in the development of solar power satellites, *Solar World Forum*, 4, 3095-3103. Pergamon Press, Oxford, 1982.

(44) Kassing, D., A survey of technologies required in the development of solar power satellites, *Solar World Forum*, 4, 3086-3094, Pergamon Press, Oxford, 1982.

(45) Kettani, M. A. and Gonsalves, L. M., Heliohydroelectric power generation, *Solar Energy* 14, 1972.

(46) Kettani, M. A., Climatological factors on heliohydroelectric power generation, Paper E38, Conf. The Sun in the Service of Mankind, UNESCO, Paris, 1973.

(47) Kettani, M. A., Solar energy activity in Saudi Arabia, Description of the Solar Energy R. & D. programs in many nations, US-ERDA, Contract E(04-3)-1122, February 1976.

(48) Abdelfattah, A., The Qattara hydrosolar project, *Solar World Forum*, 4, 3062-3069, Pergamon Press, Oxford, 1982.

(49) Abbot, C. G., The sun and the welfare of man, Smithsonian Institution, New York, 1929.

(50) Ghai, M. L., Bansal, T. D. and Kaul, B. N., Design of reflector-type direct solar cookers, *Journal of Scientific and Industrial Research* 12A, 165-175, 1953.

(51) Ghai, M. L., Pandher, B. S. and Harikishandass, Manufacture of reflector-type direct solar cooker, *Journal of Scientific and Industrial Research* 13A, 212-216, 1954.

(52) Duffie, J. A., Lappala, R. P. and Lof, G. O. G., Plastics in solar stoves, *Modern Plastics*, November 1957.

(53) Alward, R., Lawand, T. A. and Hopley, P., Description of a large scale solar steam cooker in Haiti, Paper E46, Conf. The Sun in the Service of Mankind, UNESCO, Paris, 1973.

(54) Whillier, A., How to make a solar steam cooker, Brace Research Institute, McGill University, Do-it-yourself Leaflet L2, January 1965.

(55) Walton, Jr., J. D., Development of a solar cooker using the spiral fresnel concentrator, *Solar World Forum*, 1, 596-600, Pergamon Press, Oxford, 1982.

(56) Brattle, L. V. and Irving, R. J., Energy needs for cooking in the Sudan - an interdisciplinary approach to the domestic fuel problem, *Solar World Forum*, 1, 589-595, Pergamon Press, Oxford, 1982.

(57) Spitzberg, L. A. and Williams, J. K., A linear solar concentrator system, ISES Congress, Los Angeles, Extended Abstracts, Paper 51/8, 1975.

(58) Northrup, L. L. and O'Neill, M. J., A practical concentrating solar energy collector, Paper 51/7, *Ibid*.

(59) Nelson, D. T., Evans, D. L. and Bansal, R. K., Linear fresnel lens concentrators, Paper 51/5, *Ibid.*

(60) Harmon, S., Solar optical analysis of a mass-produced plastic circular fresnel lens, Paper 51/11, *Ibid.*

(61) Löf, G. O. G. and Tybout, R. A., The design and cost of optimized systems for residential heating and cooling by solar energy, *Solar Energy* 16, 9-18, 1974.

(62) Teagan, W. P. and Sargent, S. L., A solar-powered combined heating and cooling system, Paper EH-94, Conf. The Sun in the Service of Mankind, UNESCO, Paris, 1973.

(63) Prigmore, D. and Barber, R., Cooling with the sun's heat, *Solar Energy* 17, 185-192, 1975.

(64) Swartman, R. K., Vinh Ha and Newton, A. J., Review of solar-powered refrigeration, Paper 73-WA-SOL-6, ASME, 1974.

(65) Swartman, R. K., A combined solar heating/cooling system, ISES Congress, Los Angeles, Extended Abstracts, Paper 44/8, July 1975.

(66) Wilbur, P. J. and Mitchell, C. E., Solar absorption air conditioning alternatives, *Solar Energy* 17, 193-199, 1975.

(67) Ward, D. S., Solar absorption cooling feasibility, *Solar Energy* 22, 259-268, 1979.

(68) Ishibashi, T., The result of cooling operation of Yazaki experimental Solar House 'One', *Solar Energy* 21, 11-16, 1978.

(69) Farber, E. A., Morrison, C. A., Ingley, H. A., Clark, J. A. and Suarez, E., Solar operation of ammonia/water air conditioner, ISES Congress, Los Angeles, Extended Abstracts, Paper 44/7, July 1975.

(70) Satcunanathan, S. and Kochhar, G. S., Optimum operating conditions of ammonia-water absorption systems for flat plate solar collector temperatures, Paper 44/4, *Ibid.*

(71) Alizadeh, S., Bahar, F. and Geoola, F., Design and optimisation of an absorption refrigeration system operated by solar energy, *Solar Energy* 22, 149-154, 1979.

(72) McVeigh, J. C. and Sharfi, O. A. A., Solar Powered Refrigeration: Small-scale Space Cooling in Sudan. Conf. C28, Solar Energy for Developing Countries. UK ISES, London, 1982.

(73) Ward, D. S., Duff, W. S., Ward, J. C. and Löf, G. O. G., Integration of evacuated tubular solar collectors with lithium bromide absorption cooling systems, *Solar Energy* 22, 335-341, 1979.

(74) Exell, R. H. B. and Kornsakoo, S., The development of a solar-powered refrigerator for remote villages, *Solar World Forum* 2, 1049-1053. Pergamon Press, Oxford, 1982.

(75) Ahmad, Q. A., Solar ice-making and refrigeration, *Solar World Forum* 2, 1054-1058. Pergamon Press, Oxford, 1982.

(76) Collier, R. K., The analysis and simulation of an open cycle
      absorption refrigeration system, *Solar Energy* 23, 357-365, 1979.

(77) Thomason, H. E. and Thomason, H. J. L., Solar houses/heating and
      cooling progress report, *Solar Energy* 15, 27-40, 1973.

(78) Hogg, F. C., A switched bed regenerative cooling system, Proc. XIIIth
      Int. Conf. on Refrig., Washington, DC, 4, 1971.

(79) Reed, W. R. *et al.*, Use of RBR Systems in South Australian Schools,
      Aus. Refrig., Air Cond. and Heating 26, 20-27, 1972.

(80) Rush, W., Wurn, J., Wright, L. and Ashworth, R. A., A description of
      the Solar-MEC field test installation, ISES Congress, Los Angeles,
      Extended Abstracts, Paper 44/9, July 1975.

(81) Yanagimachi, M., Report on two and a half years' experimental living
      in Yanagimachi Solar House II, Proc. UN Conf. on New Sources of Energy,
      Rome, 1961.

(82) Bliss, R. W. and Bliss, M. D., Performance of an experimental system
      using solar energy for heating and night radiation for cooling, *Ibid*.

(83) Hay, H. R., Roof-, ceiling- and thermal-ponds, ISES Congress, Los
      Angeles, Extended Abstracts, Paper 41/16, July 1975.

(84) Catalanotti, S., Cuono, V., Piro, G., Ruggi, D., Silvestrini, V. and
      Troise, G., The radiative cooling of selective surfaces, *Solar Energy*,
      17, 83-90, 1975.

(85) Raldow, W. M. and Wentworth, W. E., Chemical heat pumps - a basic
      thermodynamic analysis, *Solar Energy* 23, 75-79, 1979.

(86) Gruen, D. M., Mendelsohn, M. H. and Sheft, I., Metal hydrides as
      chemical heat pumps, *Solar Energy* 21, 153-156, 1978.

(87) Swet, C. J., Solar applications of thermochemical heat pumps - progress
      and prospects, *Solar World Forum* 3, 2265-2270. Pergamon Press,
      Oxford, 1982.

(88) Tabor, H., Solar Ponds, *Solar Energy* 27, 181-194, 1981.

(89) Tabor, H., Large-area solar collectors (solar ponds) for power
      production, U.N. Conf. New Sources of Energy, Rome 1961, reprinted in
      Solar Energy 7, 189-194, 1963.

(90) Saulnier, B., Chepurniy, N., Savage, S. B. and Lawand, T. A., Field
      testing of a solar pond, ISES Congress, Los Angeles, Extended Abstracts,
      Paper 35/1, July 1975.

(91) Weinberger, H., The physics of the solar pond, *Solar Energy* 8, 45-56,
      1964.

(92) Hawlader, M. N. A. and Brinkworth, B. J., An analysis of the non-
      convecting solar pond, *Solar Energy* 27, 195-204, 1981.

(93) Assaf, G., The Dead Sea: a scheme for a solar lake, *Solar Energy* 18,
      293-299, 1976.

(94) Dickenson, W. C., Clark, A. F., Day, A. J. and Wouters, L. F., The shallow solar pond energy conversion system, *Solar Energy* 18, 3-10, 1976.

(95) Rabl, A. and Nielsen, C. E., Solar ponds for space heating, *Solar Energy* 17, 1-12, 1975.

(96) Shah, S. A., Short, T. H. and Fynn, R. P., A solar pond-assisted heat pump for greenhouses, *Solar Energy* 26, 491-496, 1981.

(97) Styris, D. L., Zaworski, R. J., Harling, O. K. and Leshuk, J., The non-convecting solar pond. Some applications and stability problem areas. US ERDA Contract No. AT(45-1)-1830 and ISES Congress, Los Angeles, Extended Abstracts, Paper 35/2, July 1975, and in *Solar Energy* 18, 245-251, 1976.

(98) Talbert, S. G., Eibling, J. A. and Lof, G. O. G., Manual on solar distillation of saline water, Office of Saline Water, US Dept of Interior, Research and Development Progress Report No. 546, 1970.

(99) United Nations Dept. of Economic and Social Affairs, Solar distillation as a means of meeting small-scale water demands, UN Sales No. E 70 II B1, 1970.

(100) Read, W. R. W., A solar still for water desalination, Report ED 9, CSIRO, Melbourne, 1965.

(101) Cooper, P. I. and Read, W. R. W., Design philosophy and operating experience for Australian solar stills, *Solar Energy* 16, 1-8, 1974.

(102) Aegean Island installs world's largest solar distillation plant, Civil Engineering and Public Works Review, 1005, September 1967.

(103) Porteous, A., Fresh water for Aldabra, *Engineering* 490, 15 May 1970.

(104) McCracken, H., Solar stills for residential use, Paper E 6, Conf. The Sun in the Service of Mankind, UNESCO, Paris, 1973.

(105) Howe, E. D. and Tleimat, B. W., Twenty years of work on solar distillation at the University of California, *Solar Energy* 16, 97-105, 1974.

(106) Garg, H. P. and Mann, H. S., Effect of climatic, operational and design parameters on the year round performance of single sloped and double sloped solar still under Indian arid zone conditions, ISES Congress, Los Angeles, Extended Abstracts, Paper 46/4, July 1975, and in *Solar Energy* 18, 159, 1976 and *Solar Energy* 20, 363, 1978.

(107) Moustafa, S. M. A., Brusewitz, G. H. and Farmer, D. M., Direct use of solar energy for water desalination, *Solar Energy* 22, 141-148, 1979.

(108) Tanaka, K., Yamashita, A. and Watanabe, K., Experimental and analytical study of the tilted wick type solar still, *Solar World Forum* 2, 1087-1091. Pergamon Press, Oxford, 1982.

(109) Akinsete, V. A. and Duru, C. U., A cheap method of improving the performance of roof type solar stills, *Solar Energy* 23, 271-272, 1979.

(110)   Rajvanshi, A. K., Effect of various dyes on solar distillation, *Solar Energy* 27, 51-65, 1981.

(111)   Crutcher, J. L., Wood, J. R., Norbedo, A. J., Cummings, A. B. and Duffy, J. P., A stand-alone seawater desalting system powered by an 8 kW ribbon photovoltaic array, *Solar World Forum* 2, 1110-1119. Pergamon Press, Oxford, 1982.

(112)   Darwich, M. A., Alamy, F. A. and El-Sharkawy, A. I., Solar operated membrane desalination processes, Paper D5:34, ISES Solar World Forum Abstracts. Pergamon Press, Oxford, 1981.

(113)   Proctor, D. and Morse, R. N., Solar Energy for the Australian food processing industry, ISES Congress, Los Angeles, Extended Abstracts, Paper 43/2, July 1975.

(114)   Niles, P. W., Carnegie, E. J., Pohl, J. G. and Cherne, J. M., Design and performance of an air collector for industrial crop dehydration, *Solar Energy* 20, 19-23, 1978.

(115)   Ozisik, M. N., Huang, B. K. and Toksoy, M., Solar grain drying, *Solar Energy* 24, 397-401, 1980.

(116)   Bolin, H. R., Stafford, A. E. and Huxsoll, C. C., Solar heated fruit dehydrator, *Solar Energy* 20, 289-291, 1978.

(117)   North-South: *A Programme for Survival* (The Brandt Commission Report). Pan Books, London, 1980.

(118)   Grainger, W., Othieno, H. and Twidell, J. W., Small scale solar crop driers for tropical village use - theory and practical experience, *Solar World Forum* 2, 989-996. Pergamon Press, Oxford, 1982.

(119)   Gregg, D. W., Aiman, W. R., Otsuki, H. H. and Thorsness, C. B., Solar coal gasification, *Solar Energy* 24, 313-321, 1980.

CHAPTER 7

SOCIAL, LEGAL AND ECONOMIC ISSUES

INTRODUCTION

Until the mid-1970s the costs of most solar systems, particularly for space
heating applications, have been considerably greater than those using
competitive energy sources. One of the results was that a comparatively
small amount of work was carried out on the economics of solar heating and
cooling systems up to that time.

The classical paper, by Lof and Tybout (1), on optimizing the collector
parameters to minimize the total annual heating costs for the particular
geographic location, meteorological data and residential characteristics,
appeared in 1970 and, in an abridged form, in 1973 (2). Their optimization
techniques included the effects of the number of glass covers on the
collector, the collector area, the thermal storage volume and the angle of
tilt of the collector. Meteorological data was based on hourly readings of
solar radiation, atmospheric temperatures, cloud cover and wind velocity.
This work was extended (3) in 1975 to include more recent cost equations and
the effects on collector production costs of the bulk purchasing of materials
and the benefits of cost reduction by performing repetitive tasks (4).

Economic studies have also been carried out to assess the impact of solar
heating and cooling on a company supplying electricity to 7.5 million people
in an area of about 130 000 km^2 (5). The Southern California Edison Company
concluded that there were ways of combining solar heating and cooling
concepts such that the use of solar energy with electrical energy was more
economical, both for the company and the customer, than the use of either
alone. A regional economic study in Tennessee (6) in the mid-1970s, showed
that solar heating can be economical under some conditions and recommended
that government incentives in the form of tax relief should be given to
encourage the use of solar space heating and cooling.

Some years later a programme was initiated by the largest network of electric
production facilities in the USA - the Tennessee Valley Authority - to
demonstrate that the installation of 10 000 domestic water heaters can
significantly effect (i) accelerated market acceptance, (ii) manufacturer's
interest, (iii) the financial community's awareness and (iv) the rate of
construction of nuclear and fossil fuelled power plants. The results from
the very wide ranging series of interviews (7) showed enormous variations in

161

attitudes and experience. The major conclusion was that the peak electrical
demand will fall. This will allow the proliferation of nuclear plant
construction to be constrained while "our alternative future planners
contemplate what to do next".

For large-scale electricity and process heat generation the impact of 'solar
thermal repowering' in the southwestern states of the USA has been analysed
(8). Repowering involves adding the solar thermal central receiver, or
power tower, as a retrofit to an existing energy system. Among the major
findings of the study were that the estimated electricity generating market
potential was 18 000 MWe, equivalent to 24% of the 1978 capacity of existing
oil and gas-fired steam turbine units, together with a possible 20% of the
demand for industrial process heat. Total system costs were projected to
be less than $1000 (at 1978 prices) per kWe. Basic comparisons of the
effects of alternative incentives on the rate of solar energy utilization
have been provided by Peterson (9), who included a consumer decision model
in his analysis. The model had to meet at least three basic requirements.
The first was that consumer behaviour is not based entirely on economic
efficiency. The second was that there should be some allowance for
differences in consumer preferences and the third that consumer behaviour may
change as increased knowledge and experience is gained with solar energy.

MARKET PENETRATION

An analysis by the International Institute for Applied Systems Analysis,
Austria, of the potential role of solar energy applications in Europe and
the USA over the coming century showed that if new solar technologies were
to provide 1% of US primary energy by 1985 and penetrated the US energy
marketplace at a fractional rate substantially greater than that of oil or
gas in the past, it would still require seventy years for solar energy to
displace 50% of other primary sources (10). This slow rate of penetration
is consistent with the requirements for and availability of investment
capital, including competition for this capital for other purposes such as
housing, agriculture and new industrial facilities. Their work suggested
that new institutional and economic mechanisms would be required to
accelerate the embedding of solar technologies in national economies if
these technologies were to make a significant contribution to energy needs
by the early part of the next century.

IMPACT ON THE ECONOMY

One of the conclusions reached in a study (11) of the effect of solar space
heating and domestic hot water systems on the new housing market during the
period 1980-2000 was that the solar approach enabled 5000 new jobs to be
created per million tonnes of domestic fuel oil saved, compared with the
creation of only 600 new jobs by nuclear generated electric heating systems.
The Institut de Sciences, Mathématiques et Economiques Appliquées in Paris
drew a careful distinction between the supply or demand factors as main
influences on economic growth. If investment is determined by supply, the
decisive factor in choosing the technical alternative is competitiveness,
often accompanied by maximum productivity. But if investment is determined
by demand with the market protected from foreign competition, the technical
choice can be oriented towards those technologies which have, at equal cost,
the highest job requirements. The different multiplier effects show that
*if one job is created by the nuclear-powered electric heating systems, 2-2.5
jobs could be created by developing solar energy systems and heat pumps.*

Among many different aspects affecting consumer attitudes which were reported
in Brighton at the Solar World Forum, the characteristics of subsidies were
examined by Feiveson and Rabl (12). They found that subsidies were of two
kinds: those proportional to the solar investment, such as tax credits, and
those proportional to energy saved. Both can have disadvantages. Subsidies
proportional to investment discriminate against technologies which are
difficult to define administratively, e.g. conservation and passive solar.
They could encourage inefficient consumer behaviour by making a householder
prefer a subsidized active solar space heating system to more effective, but
unsubsidized conservation technologies. Subsidies proportional to energy
saved could raise intractable problems of measurement. They concluded that
a rational society would not subsidize solar energy but would instead tax
each energy source in such a way that everybody pays the full social cost of
the energy he consumes.

The problems of measuring the energy saved did not appear to worry de Lima
and Seisler (13), who felt that solar energy for buildings can be
economically viable if the systems are metered, making residents financially
responsible for their consumption of solar supplied energy as well as energy
supplied by traditional fuels. Building owners or utility companies who may
wish to invest in a solar system as a hedge against rising fuel costs could
then obtain an acceptable return on their investment, and they gave some case
studies of buildings where metered solar energy systems have been installed.

EDUCATION

A major need for many types of educational programme related to solar energy
was recognised at an Educational Workshop held at the Solar World Forum in
1981. While many differences of approach from the various countries could
be seen, four main sectors where solar and other energy related topics
should be included in curriculum development, were identified as follows:

(i)   From age 5+   Schools. Starting with the simplest ideas and
      up to about 20   materials
(ii) From age 16+,   Technical College - Technician or mechanic courses
      but in most
      countries 18+   University, polytechnic, college of technology -
                             degree and postgraduate courses, e.g. engineering
                             or architecture

(iii) From 25+   Mid-career courses   technicians
                          Retraining, up-dating   installers
                                                    professional level
                                                    e.g. architects

(iv) Any age   Public awareness - national, regional and local
                             government officers, administrators, civil
                             servants, the general public.

Several approaches to this work in different countries have been described.
For example, the University of Southern California programme (14) started
with the youngest age group (Elementary Schools) and spread out the work,
with simple laboratory demonstrations to cover all types of school. Over
2000 copies of their Elementary School curriculum were distributed to 40
states in the USA. Community based 'solar action' programmes including
demonstrations, seminars and tours of solar buildings have been supported by
grants from the US Department of Energy (15). A possible approach to
training in solar technologies in the developing countries has been described
by Deb (16) who pointed out that a major appeal of solar energy for these

countries is its adaptability to decentralised applications, preferably
based on intermediate technologies.

THE RIGHT TO SUNSHINE

The problem which could face anyone who has installed a solar system of any
type, active, passive, hybrid or photovoltaic, is simply stated but difficult
to solve. What can be done if a neighbour builds a new dwelling or extends
an existing building or allows a bank of trees to grow on his land, thereby
shading your collectors? In an excellent analysis containing 17 references
to this issue, Eisenstadt and Utton (17) point out that at least three
pertinent questions arise:

  (i)   does the homeowner have a right to the sunshine that is blocked by
        his neighbour's building or tree?
  (ii) if he does not have a right to the solar energy, should such a right
        be given him?
 (iii) if such a right is given, how should it be done?

The answers to these questions will naturally vary from one country to
another. In English common law the doctrine of ancient lights stated, as
far back as the seventeenth century, that if a person had the uninterrupted
use of light and air through a window for 20 years an adjoining landowner
could not cause the light to be blocked. This doctrine came to the American
colonies, but during the nineteenth century it was gradually rejected.
During the twentieth century the case of an extension to one hotel shading
the swimming pool of the other after 2 p.m. resulted in the court stating:

   No American decision has been cited, and independent research has
   revealed none, in which it has been held that - in the absence of some
   contractual or statutory obligation - a landowner has a legal right to
   the free flow of light and air across the adjoining land of his
   neighbor. (18)

The answer to the second question must be positive. The right to solar
energy would encourage many more practical applications. But, as Eisenstadt
and Utton (17) comment:

   The problem will be exacerbated when the first lawsuit involving solar
   collector shading is decided in favor of the party doing the shading.

They also discussed methods for creating solar rights in the USA and gave
examples to illustrate how the need of one party for solar energy should be
balanced against the inconvenience to his neighbour.

ECONOMIC ANALYSIS

The two aspects in the economic assessment of solar systems, the financial
and the proportion of the total energy demand which can be met by solar
energy, are integrated in any complete assessment. A separate analysis of
various alternative approaches is given in the following sections.

NOTATION

C The total capital cost of the solar heating system, including installation, ancillary equipment and collector panels, (£).

C_m The total capital cost per m^2 of collector area.

f The annual inflation rate in the price of competitive energy, expressed as a fraction so that $100r$ is the % annual inflation rate.

F Annual savings resulting from the substitution of the solar system for a proportion of the competitive energy system, (£).

F_C The cost of competitive energy, (£/kWh).

G_i The total annual irradiation incident on the collectors for a given orientation and angle of tilt (kWh/m^2).

i The annual net effective interest rate, defined as $\dfrac{(1 + r)}{(1 + f)} - 1$. $100i$ is the % annual net effective interest rate.

ℓ A 'loss of efficiency' factor used in Table 7.2.

M The capital repayment factor or amortization rate, defined as Y/C with no inflation and Z/C with inflation. $100M$ is the % capital repayment factor.

n The period of years over which the total capital cost is repaid (or the estimated lifetime of the solar heating system).

r The annual interest rate on borrowed capital. $100r$ is the % annual interest rate.

T A maintenance cost (£), used in Table 7.2.

Y The constant annual payment necessary to repay a capital loan of £C in n years (£/annum).

Z The annual payment originally considered necessary when the capital loan commenced. With inflation, the first payment becomes $Z(1 + f)$.

η_c The efficiency of the collectors, expressed as a fraction.

η_s A system efficiency factor, which ideally is unity, but is less in practice as it allows for losses and differences between individual installations.

STANDARD PRESENT VALUE ANALYSIS

Conventional discounted cash flow concepts are used to establish the present value of future savings and can be used to determine if it is economically justifiable to invest in a solar heating system, based on the anticipated savings in heating costs over the lifetime of the system.

The present value of £1 needed in n years' time, is obtained by assuming today's money is invested at the 'market discount rate' d.

Present value $\times (1 + r)^n$ = £1 in n years' time or present value = $1/(1 + r)^n$. An analysis of appropriate discount rates and the use of payback periods in assessing the economics of solar heating has been given by Sulock (19). Even with the high interest rates prevailing in the UK during the late 1970s, inflation rates were consistently higher than the rates at which money could be borrowed for solar installations, as these were regarded as an 'improvement' and the householder could get tax relief or an addition to his mortgage.

The equation used to compute the constant annual payment Y necessary to repay a capital loan C in n years at a fixed annual interest rate r is

$$Y = \frac{Cr(1 + r)^n}{(1 + r)^n - 1} . \tag{7.1}$$

This equation is derived by considering the outstanding balance left at the end of each year when the interest and a proportion of the initial loan have been repaid. At the end of the first year the outstanding balance is

$$C(1 + r) - Y. \tag{7.2}$$

At the end of the mth year, where m is any number between 1 and n, the outstanding balance can be expressed as

$$C(1 + r)^m - \frac{Y}{r}(1 + r)^m + \frac{Y}{r} . \tag{7.3}$$

At the end of the nth year the outstanding balance is zero, and by substituting n for m in equation (7.3) and equating to zero, the final expression becomes

$$C(1 + r)^n = \frac{Y}{r}(1 + r)^n - \frac{Y}{r} \tag{7.4}$$

which is the same as equation (7.1).

THE EFFECT OF INFLATION

The greater the rate of inflation, the more worthwhile it becomes to make a capital investment now to reduce recurrent costs at later dates. Using the same procedure as in the previous section, with Z as the annual payment originally considered necessary and f the annual inflation rate, the outstanding balance at the end of the first year is

$$C(1 + r) - Z(1 + f). \tag{7.5}$$

At the end of the second year the outstanding balance is

$$\{C(1 + r) - Z(1 + f)\}(1 + r) - Z(1 + f)^2$$

which can be rewritten as

$$C(1 + r)^2 - Z(1 + f)^2 \left(1 + \frac{(1 + r)}{(1 + f)} \right) \tag{7.6}$$

and at the end of the nth year the outstanding balance is zero and the final
expression becomes

$$C(1 + r)^n = Z(1 + f)^n \left(1 + \frac{(1 + r)}{(1 + f)} + \cdots \frac{(1 + r)^{n-1}}{(1 + f)^{n-1}} \right)$$

which can be written as

$$C \frac{(1 + r)^n}{(1 + f)^n} = Z \left(\frac{\frac{(1 + r)^n}{(1 + f)^n} - 1}{\frac{(1 + r)}{(1 + f)} - 1} \right). \tag{7.7}$$

The annual net effective interest rate, i, can now be introduced, where
$(1 + i) = \frac{(1 + r)}{(1 + f)}$ or $i = \frac{(1 + r)}{(1 + f)} - 1$ and equation (7.7) becomes

$$Z = \frac{Ci(1 + i)^n}{(1 + i)^n - 1} \tag{7.8}$$

which has a similar form to equation (7.1).

Over the next few decades the gradual depletion of the world's fossil fuel
reserves will mean that fossil fuel costs must inevitably continue to
increase. The rate of increase may be less than that experienced from 1974
to 1976, so that up to 15% annually could be reasonable anticipated. This
will tend to keep the net effective interest rate fairly close to zero and
will have a very considerable influence on the capital investment in a solar
heating system which can be economically justified. Fig. 7.1 shows the
relationship between the capital repayment factor M, defined as Z/C, and the
net effective interest rate for different repayment periods. For economic
viability the capital repayment factor M should be less than F/C, or the
total capital cost of the system should not exceed F/M. The use of equation
(7.8) or Fig. 7.1 is illustrated in the following example:

The amount of heat which can be supplied by a solar heating system is
8000 kWh/annum. Competitive energy costs are £0.05/kWh. The anticipated
increase in competitive energy cost is 14% per annum and the interest
rate on a loan for the installation is 11.5% per annum. How much could
be invested in the solar heating system if the loan is to be repaid in
12 years?

The effective interest rate $i = \left(\frac{1.115}{1.14} - 1 \right) = -0.02193$ hence $M = 0.0719$
(from equation (7.8) or Fig. 7.1).

The capital which can be invested should be less than $\frac{0.05 \times 8000}{0.0719} = £5563$.

Figure 7.1 also shows that investment in solar equipment is essentially a
long-term investment, as capital repayment factors smaller than 0.1 can only
be achieved with a ten year period if the inflation rate is greater than the
interest rate. Values for a net effective interest rate of zero, obtained
when $i = r$, are sometimes known as the payback period.

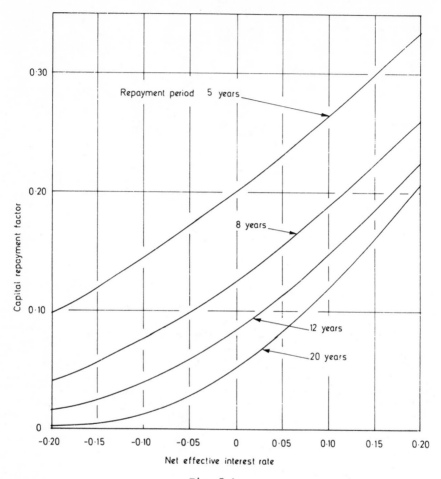

Fig. 7.1.

MARGINAL ANALYSIS

The capital repayment factor can be used to carry out a marginal analysis on any system to examine the effect of variations in collector design or system parameters, an approach which has been used by Winegarner (20). Marginal analysis can help to decide which collector should be used in a given system, or whether a selective surface or double glazing is worthwhile. Equation (7.8) can be expressed in terms of the total capital cost per unit area of collector, C_m, as follows:

$$C_m = \frac{F_c G_i \eta_c \eta_s}{M} \qquad (7.9)$$

where F_c is the cost of competitive energy, G_i the incident radiation on the collectors, η_c the collector efficiency and η_s a system efficiency factor. Differentiating with respect to collector efficiency gives

$$\Delta C_m = \frac{(\eta_s F_c G_i)}{M} \Delta \eta_c \qquad (7.10)$$

The use of this equation can be illustrated by an example:

The application of a certain selective surface can improve the overall efficiency of a collector from 47% to 53%. The net effective interest rate is +0.03 for a period of eight years. The cost of competitive energy is £0.05/kWh, G_i is 930 kWh/m²/annum and η_s is 0.73. What is the maximum allowable additional cost of the selective surface?

From equation (7.8), $M = 0.1425$
or Fig. 7.1.

$$\Delta \eta_c = 53\% - 47\% = 0.06$$

$$\therefore \quad \Delta C_m = \frac{0.73 \times 0.05 \times 930 \times 0.06}{0.1425} = £14.29/m^2.$$

The maximum allowable additional cost for the selective surface is £14.29/m² and a commercial decision can be made about the manufacturing costs.

OPTIMIZING THE COLLECTOR AREA

A full treatment of this topic is beyond the scope of this book and a very comprehensive treatment of the solar collector system and the amount of energy which could be collected over a year has been given by Duffie and Beckman (21). Lof and Tybout (1, 2) established that two parameters had practically no effect on the system performance, the heat transfer coefficient of the insulation on the storage tank and the heat capacity of the collector. The optimum tilt of the collector was found to be between ten and twenty degrees greater than latitude for sites of widely different latitude, and variations in collector tilt from latitude to latitude plus thirty degrees made very little difference to costs. Larger collector areas which are needed in colder climates should be accompanied by larger storage volumes.

The optimizing procedure commences by calculating the output from the solar collectors, based on hourly data if possible, which is then compared with the demand. In winter the total demand will probably not be met, while in the summer more heat is available than is needed. The relationship between the percentage of the total demand which can be met by solar energy and collector area/floor area ratio follows a curve broadly similar to that shown in Fig. 7.2. The capital costs of a solar heating system are also shown in Fig. 7.2. There is a fixed initial sum for the pump, storage tank and control system, while additional collectors are assumed to be at a fixed unit cost. For each system under consideration a table can be drawn up giving the total annual running costs, based on the discounted solar system cost and the proportionate costs of the auxiliary system. This is used to plot a curve relating the total cost to the collector area, from which the optimum area can be determined.

The procedure is best illustrated by an example where the annual demand for a floor area of 100 m² is assumed to be 10 000 kWh. Figure 7.2 gives the relationship between collector area and percentage demand met. The capital cost of the solar heating system is assumed to be £400 plus £70/m² of collector area. The repayment period is twenty years at a net effective

interest rate of +5%, which gives a capital repayment factor of 0.08024. A
table can now be drawn up in which the solar cost column is obtained from
the product of the capital repayment factor and £(400 + 70A), where A is the
area of the collectors. The percentage demand met by auxiliary heating is
obtained by subtraction and, with an assumed cost of £0.05/kWh for auxiliary
heating, the total annual heating costs are obtained by adding the solar and
auxiliary costs as shown in Table 7.1.

Table 7.1.

Collector area (m^2)	Solar cost (£)	% Demand met by solar	% Demand met by auxiliary	Auxiliary cost (£)	Total cost (£)
10	88.26	34	66	330	418.3
20	144.44	50	50	250	394.4
30	200.6	62	38	190	390.6
40	256.8	70	30	150	406.8

The total can be plotted against collector area to show that the optimum
collector area for the system is approximately 30 m^2. Accurate predictions
of the optimum collector area and the effect of the other main parameters
are best obtained by a computer analysis which can use the hourly
meteorological data.

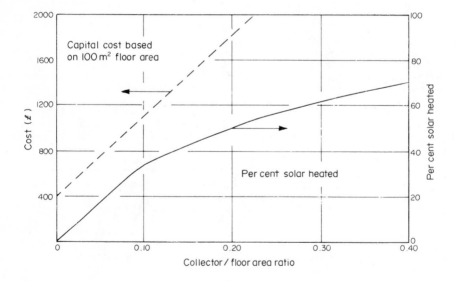

Fig. 7.2.

VARIABLE INTEREST AND INFLATION RATES

Interest rates and inflation rates will not remain constant for long periods.
The performance of a solar collector may deteriorate and the system may need
maintenance. All these factors can be assessed by an extension of the step-
by-step procedure outlined in the derivation of equations 7.1 and 7.8. A
tabular method is used in which F is calculated using conventional
techniques, but an allowance can be made for a gradual reduction of $n_c\ n_s$ by
multiplying F by an 'efficiency loss' factor ℓ. The value of F tends to
increase from year to year due to inflation, but this may be offset by a
deterioration in collector performance. The method is illustrated in the
following example, which is based on the figures for an 8.17 m^2 domestic
heating system which shared the first prize in the 1975 Copper Development
Association Solar Heating Competition (22). The actual total cost of the
system was estimated to be £539, but when an allowance was made for savings
in the original roof, this reduced to £367. The efficiency loss factor ℓ
decreases by 0.02 for five years, then a full maintenance, T, costing £80
restores ℓ to 1.00. Thereafter it continues to decrease at 0.02 per annum.
The competitive energy costs are assumed to be constant for the first year,
then increase by 10% annually for three years, then by 5% annually.

Table 7.2

Year	C	r	$(1 + r)C$	i	F	ℓ	T
1	376.00	8	406.80	0	63.00	1.00	0
2	343.08	8	370.52	10	69.30	0.98	0
3	302.61	9	329.84	10	76.23	0.96	0
4	256.66	10	282.33	10	83.85	0.94	0
5	203.51	8	219.79	5	88.04	0.92	80
6	218.79	7	234.11	5	92.44	1.00	0
7	141.67	7	151.58	5	97.06	0.98	0
8	56.46	9	61.54	5	101.92	0.96	0

In year 3, for example, $1.10 \times 69.3 = 76.23$ and (302.61×1.09) −
$(76.23 \times 0.96) = 256.66$, the new C for year 4. A credit of £36.3 is left at
the end of the eighth year.

One of the really interesting features of this detailed analysis is that
during the early years of a solar installation it can show that the amount
of money saved annually by the solar heating system does not have to be
greater than the annual interest on the capital sum borrowed.

It is also important to appreciate that the value of the currency unit, in
this case the pound (£), falls each year with inflation. This is illustrated
by the following extract from the Digest of UK Energy Statistics.

General index of retail prices in the UK.
(Monthly average 15 Jan 1974 = 100)

	1970	1973	1979	1980
All items	73.1	93.5	223.5	263.7
Fuel and Light	77.3	94.5	250.5	313.2
Coal and smokeless fuels	74.0	95.0	265.1	338.3 ·
Gas	85.0	99.0	182.5	212.8
Electricity	78.0	94.0	282.6	359.5

It can be calculated that the cost of electricity was increasing at a real
rate of about 6% per annum relative to the general index of retail prices
from 1973 to 1980. This reinforces the argument in favour of an investment
in a solar system.

Any solar heating system for domestic hot water, space heating or industrial
applications, will save energy. In the long term this consideration may
outweigh all other factors.

REFERENCES

(1) Löf, G. O. G. and Tybout, R. A., Solar house heating, *Natural Resources
     Journal* 10, 268, 1970.

(2) Löf, G. O. G. and Tybout, R. A., Cost of house heating with solar
     energy, *Solar Energy* 14, 253-278, 1973.

(3) Pogany, D., Ward, D. S. and Lof, G. O. G., The economics of solar
     heating and cooling systems, ISES Congress, Los Angeles, Extended
     Abstracts, Paper 12/1, July 1975.

(4) Behrens, H. J., The learning curve, from *Cost and Optimization
     Engineering*, F. C. Jelen, Ed., McGraw-Hill, New York, 1972.

(5) Braun, G. W., Davis, E. S., French, R. L. and Hirshberg, A. S.,
     Assessment of solar heating and cooling for an electric utility
     company, ISES Congress, Los Angeles, Extended Abstracts, Paper 12/5,
     July 1975.

(6) Lunsdaine, E., Reid, R. L. and Albrecht, L., Regional economic study of
     solar heating (Tennessee), ISES Congress, Los Angeles, Extended
     Abstracts, Paper 12/2, July 1975, and in *Solar Energy* 19, 513-517,
     1977.

(7) Talbot, J. G. and Wiener, M., Solar Nashville: the effect of 10 000
     installed DHW units on market acceptance, Paper J1:13, ISES Solar
     World Forum Abstracts, Pergamon Press, Oxford, 1981.

(8) Lord, N., Curto, P. and True, S., Solar Thermal Repowering, MITRE
     Technical Report MTR-7919, The Mitre Corporation, 1820 Dolley Madison
     Boulevard, McLean, VA 22102, USA, December 1978.

(9) Peterson, H. C., The impact of tax incentives and auxiliary fuel prices
      on the utilization rate of solar energy space conditioning, NFS-RANN
      Grant Nos. AER-09043-A01 and APR 75-18004, Utah State University,
      Logan, Utah, 1976.

(10) Nakicenovic, N. and Weingart, J. M., Market penetration dynamics and
      constraints on the large-scale use of solar energy technologies, Paper
      J1:3, ISES Solar World Forum Abstracts, Pergamon Press, Oxford, 1981.

(11) Outrequin, P., Solar energy and economic growth, *Solar World Forum*
      2, 1395-1403, Pergamon Press, Oxford, 1982.

(12) Feiveson, H. A. and Rabl, A., Characteristics of solar subsidies,
      Paper J2:32, ISES Solar World Forum Abstracts, Pergamon Press, Oxford,
      1981.

(13) de Lima, H. and Seisler, J. M., Metering the sun: the cost of solar
      energy to consumers, Paper J2:33, ISES Solar World Forum Abstracts,
      Pergamon Press, Oxford, 1981.

(14) Lampert, S., Wulf, K. and Yanow, G., Educational programs in solar
      energy, *Solar World Forum* 2, 1556-1567, Pergamon Press, Oxford,
      1982.

(15) George, K. L. and Salmon, G. L., Solar commercialization on a community
      level of the solar action program, Paper J4:50, ISES Solar World Forum
      Abstracts, Pergamon Press, Oxford, 1981.

(16) Deb, S., A possible pattern of training courses in heliotechnology for
      developing countries, Paper J4:48, ISES Solar World Forum Abstracts,
      Pergamon Press, Oxford, 1981.

(17) Eisenstadt, M. M. and Utton, A. E., On the right to sunshine, *Solar
      Energy* 20, 87-88, 1978.

(18) *Ibid.*, Ref. 4.

(19) Sulock, J. M., The economics of solar heating. A note of the
      appropriate discount rate and the use of the pay back period, *Solar
      Energy* 24, 585-586, 1980.

(20) Winegarner, R. M., Coatings, Costs and Project Independence, *Optical
      Spectra,* June 1975.

(21) Duffie, J. A. and Beckman, W. A., *Solar Engineering of Thermal
      Processes,* John Wiley, New York, 1980.

(22) Copper Development Association, Orchard House, Mutton Lane, Potters
      Bar, Hertfordshire EN6 3AP, UK.

CHAPTER 8

PHOTOVOLTAIC CELLS, BIOLOGICAL CONVERSION SYSTEMS AND PHOTOCHEMISTRY

PHOTOVOLTAIC CELLS

The direct conversion of solar energy into electrical energy has been studied since the end of the nineteenth century. The early work was concerned with thermocouples of various different alloys, and efficiencies were very low, usually less than 1%. This work was reviewed by Telkes (1) in a paper written in 1953 and at that time it was felt that little more could be achieved and that efficiencies in this order were quite unsuitable for the generation of electricity. A similar pessimistic view was expressed in the UK (2). However, in 1954 the Bell Telephone Laboratories in the USA discovered that thin slices of silicon, when doped with certain traces of impurities became a factor of ten times or more efficient at the conversion of solar radiation to electricity than the traditional light sensitive materials used in earlier photocells. Since then there has been a steady history of improvement and considerably higher conversion efficiencies have been quoted - up to 19% for silicon cells and over 20% for certain new gallium arsenide cells under laboratory conditions. The use of solar cells in space applications is well known and the development for terrestrial applications has accelerated so rapidly over the past few years that it is anticipated that solar cells will be providing a significant proportion of the electrical energy requirements of many countries throughout the world by the end of the century.

Among the advantages listed for the modern solar cell are that it has no moving parts to wear out, has an indefinitely long life, requires little or no maintenance and is non-polluting (3). Unlike other types of electrical generator it is suitable for a wide range of power applications from less than a watt to several thousand megawatts.

Although a Japanese estimate considered that a 10 MW generating station made with the technology available in 1974 would require the entire world production of silicon, about 1000 tonnes at that time, more recent estimates in 1981 (4, 5) suggest that amorphous silicon, cast polycrystalline silicon and cadmium sulphide are free of serious problems of material supply. The Japanese 'Project Sunshine', which started in 1974 when the estimated total photovoltaic power being generated there was some 20 kW, included plans for a 100 MW system by 1991, with even larger plants operating by 2000. The

National Science Foundation in the USA (6) estimated that by 1990 there should be 5000 MW at peak output manufactured annually in solar arrays and 20 000 MW at peak output manufactured by 2000, which would be approximately 2% of the projected total electrical power demand.

Types of Solar Cell

The doping of a very pure semiconductor with small traces of impurities can modify its electrical properties, producing two basic types: p-type, having fixed negative and free positive charges, and n-type, having fixed positive and free negative charges. If these two types are placed together and the surface is exposed to sunlight, electrons will diffuse through the p-n junction in opposite directions giving rise to an electric current, as shown in Fig. 8.1. The earliest solar cells were made of silicon and one type of modern silicon cell is made by doping a slice cut from a single crystal of highly purified silicon with phosphorous, arsenic or antimony and diffusing boron into the upper surface, forming a 'p-on-n' cell.

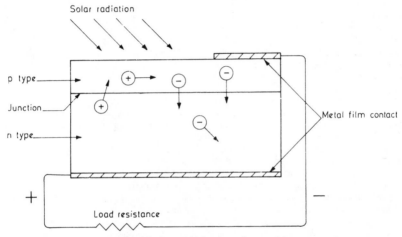

Fig. 8.1.   Silicon solar cell.

An alternative method, forming a 'n-on-p' cell has proved to be more effective and is preferred. A detailed discussion of the physics of solar cells is beyond the scope of this text, but is widely available in the literature (7-10). The front of the cell is protected by a transparent cover. This was usually made of thin glass or fused silica (quartz) for the early space applications, but a wide variety of other methods using weather resistant plastics had been adopted for many terrestrial applications by 1980 (10).

The commercial production processes in the mid to late 1970s were complex. For example, during one stage, the 'pulling' of a crystal from the melt, temperature control within $\pm 0.1^{\circ}C$ at $1420^{\circ}C$ was involved (11). Consequently these early solar cells were expensive. When assembled in a group of encapsulated cells, known as modules, module costs were between US$25-90 per peak watt[*] in 1975, expressed in 1975 dollars. The production of these very

---

[*]A peak watt (Wp) is defined as one watt output at a light intensity of 1000 W/m^2 and a temperature of $25^{\circ}C$.

pure single silicon crystals, or electronic grade silicon, was a major
factor in this high cost and it was hoped that a solar grade silicon could
be developed. This would have more impurities than the electronic grade
silicon, but would perform as well at less than one fifth of the cost. By
1981, however, this breakthrough had not been achieved and many other
processes were under development. Amorphous, or uncrystallized, silicon can
be deposited in very thin films or layers. This could lead to relatively
inexpensive cells in the small power range, typically less than one peak
watt, where their relatively low efficiency, in the order of 3-5%, would be
of minor importance. Polycrystalline silicon, which consists of many small
single crystals arranged at random, has also been deposited in thin layers on
a wide variety of materials by different methods such as fusion, dipping or
coating. Practically all these processes give cells with laboratory
efficiencies higher than 8% (12), with some greater than 15%. A radically
different production technique has been investigated in several laboratories.
By growing silicon crystals directly in ribbon form it is possible to
eliminate the present costly process of cutting very thin wafers of silicon
from large cylinders of single crystal. One process, known as edge defined
film-fed growth (EFG) was first investigated at the Tyco laboratories (13).
It was applied to silicon ribbon in 1971 and the first objective was to grow
silicon ribbons suitable for use in solar cells. The second objective was
to produce high quality silicon ribbon so that approximately 10% efficient
solar cells could be achieved and the third objective was to grow the ribbon
in a continuous length to indicate the basic possibility of using the
technique for production of very long ribbons.

These objectives have been achieved, with lengths of nearly 2 m approached in
1974, and a prototype multiple-ribbon production unit producing five 50 mm
wide ribbons at a rate of about 40 mm per minute by 1979. Another, the edge-
stabilized ribbon (ESR) process developed by Arthur D. Little Inc. in 1981
as a result of research work carried out at Massachusetts Institute of
Technology by Emanual Sachs, was considered to represent "a significant
milestone in the development of photovoltaic materials to meet industry
production goals" (14). In the first results announced, ribbon lengths of
up to 4 m and 25 mm wide, had been obtained at manufacturing rates from 25
to 150 mm per minute, with the possibility of continuously adjusting the
thickness from 4 μm to well over 300 μm. Widths of over 100 mm were
considered to be technically feasible. An important advantage over other
ribbon growth methods which need the temperature to be controlled within
one-tenth of a degree, is that the ESR process can perform reliably with
temperature variations greater than $10^{\circ}C$.

A wide variety of other materials in addition to silicon are being
considered for use in solar cells. The photovoltaic effect in single
crystal cadmium sulphide (CdS) and in thin films of cadmium sulphide was
discovered in 1954 and by the mid-1970s it was considered that the necessary
processes for the mass production of the low-cost cadmium sulphide solar
panel array had already been developed. However, as Rauschenbach (10)
subsequently pointed out, every year a group of workers announced the
solution to the (problems of) self-degradation and instability mechanisms
of these cells. By the early 1980s a lasting solution had still to be
found, although laboratory efficiencies had been steadily increasing to
about two-thirds of the theoretical maximum of 15%. Life expectancies of
over twenty years are predicted from accelerated life tests.

Two types of cell with fairly similar properties are gallium arsenide and
indium phosphide. Gallium arsenide has the ability to withstand considerable concentration and up to 1000 times full sunlight has been reported by

the Plessey Company in the UK and Varian Associates in California. Research is also being sponsored into organic semiconductors and Schottky barriers (metal-to-semiconductor junctions). A more detailed description of this work has been given by Jesch (15).

Applications

Although the present high cost of commercially available solar cells makes them unattractive in any situation where a conventional electricity supply is available, there are an increasing number of applications where solar cells are already economically competitive. One of the earliest applications was in the provision of unmanned lights at sea. Flashing lights on buoys, lighthouses and offshore oil rigs are increasingly being powered by solar cells, particularly in the Gulf of Mexico and the off-shore islands in Japan (16). An early application in the UK was the installation of a silicon cell array to power a navigation light at Crossness on the River Thames in 1968 (17). The main problem in earlier tests was the hostile marine environment and the salt in the atmosphere was found to attack some resins and plastic based mounting boards.

Automatic weather stations and other remote instruments that are difficult to reach are now being considered for solar powered operation. Although the total costs using conventional fossil fuels may appear to be lower, they must also be offset against the cost of access for maintenance and refuelling. One of the earliest uses in the USA was for powering remote radio transmitters on the tops of mountains for the US Forest Service. In Nigeria the school television programmes are intended for schools located in regions not provided with electrical power and solar cells have been used to power television receivers since 1968 (18). The power consumption is about 32 watts from a d.c. power supply of between 30 and 36 volts. The solar cells had an initial cost of $3100 compared with $976 for chemical batteries, but the estimated lifetime of the solar cells, about 10 years, would give over 25 000 hours' operation, compared with the 2000 hours for the chemical batteries.

Among other early applications reported by Centralab (16) were the powering of the world's first highway callbox system in California, the use of tiny radio transmitters attached to migrating wild animals, remote snow and water gauges, and fire alarms and seismographs.

By 1975 the world's largest terrestrial photovoltaic system was the solar array of 1 kW peak capacity installed by a commercial company in the USA, the Mitre Corporation (19). Their energy storage system consisted of battery storage for short-term and peak power requirements, and an electrolysis hydrogen gas system fuel cell combination for base load and operation at night.

The next six years saw a remarkable three-hundredfold increase in the maximum size of the solar array, with a 25 kW water-pumping system in Nebraska by 1977, a 60 kW system augmenting a grid system in California in 1979 and two systems of over 300 kW reported during 1981. The first was a concentrating collector photovoltaic system to provide the Mississippi Community College in Arkansas (20) with both electricity and thermal energy. The peak electrical output was designed to be 320 kW from 59 400 single-crystal

Fig. 8.2.

silicon cells, with parabolic trough concentrating collectors producing an
average concentration of 30 suns by tracking the sun throughout the day.
The 270 concentrator units occupy an area of about 1.5 ha, with a collector
surface area of 3500 m^2 capable of producing about 7000 kWh thermal per day
for winter heating. The second was due for completion in 1982 when it
would be the world's largest system with a 350 kW peak output and would
supply the villages of Al-Aineh and Al-Jubaila in Saudi Arabia. The project
formed part of the $100 million five-year joint programme between Saudi
Arabia and the USA, known as SOLERAS. Other SOLERAS projects included
several types of air-conditioning coolers and solar-powered desalination
units. In addition to the development of these larger systems, there are
a number of applications, particularly in the medical field, where there
is a need for a smaller, secure electricity supply. A good example is the
photovoltaic array supplying the San Hospital, Mali, shown in Fig. 8.2.

During 1981 the Sacramento Municipal Utility District (SMUD) confirmed that
the United States Department of Energy had been asked to underwrite a
proposal to build a 100 MW photovoltaic power plant at the Rancho Seco
nuclear power plant near Sacramento. It is to be built in stages over a
period of twelve years with the first stage of 1 MW due for completion after
18 months' work. The final stage would be 35 MW, and when completed would

show a further three-hundredfold increase in size from 1982. System costs
in the final stage were predicted to be about $2.16 per peak watt (in 1981
prices) a figure as low as estimates given for nuclear power installations
in California, but about twice as high as the figures suggested for targets
by the United States Department of Energy. The effects of inflation and
fluctuating exchange rates make it difficult to assess any predictions of
future costs, but it is significant that the 1975 target of less than $2 per
peak watt by 1985 will very probably be achieved as the following table,
based on the United States Department of Energy price goals and prepared in
1980 dollars, indicates (4).

Table 8.1

Year	Module price ($ per Wp)	System price ($ per Wp)	Prospective market
1986	0.70	1.60-2.20	Utility-connected residences and intermediate load centres
1990	0.15-0.40	1.10-1.30	Central power stations

System costs include the array structure, wiring, controls and energy
storage as well as the land, fencing and installation.

The long-term future has been aptly summarized by Dr H. L. Durand, President
of the Commissariat à l'Energie Solaire in France (12)

> If photovoltaic modules remain above one US dollar [1979 dollars] per
> peak watt, a large industry can be built in order to replace the
> billion-dollar market of small and medium size power stations, a market
> which is held today by diesel generators. If the cost comes down below
> 0.5 US dollars/Wp the jump may become fantastic, since photovoltaics
> may then compete with the large fuel or nuclear stations, provided that
> the storage problem can be solved . . . The hopes certainly justify the
> large amount of effort and money which are currently allocated to this
> exciting challenge.

BIOLOGICAL CONVERSION SYSTEMS

Solar energy can be used by all types of plants to synthesise organic
compounds from inorganic raw materials. This is the process of
photosynthesis. In the process carbon dioxide from the air combines with
water in the presence of a chloroplast to form carbohydrates and oxygen.
This can be expressed in the following equation:

$$CO_2 + H_2O \xrightarrow[\text{chloroplast}]{\text{sunlight}} C_x(H_2O)_y + O_2. \qquad (8.1)$$

A chloroplast contains chlorophyll, the green colouring matter of plants.
The carbohydrates may be sugars such as cane or beet, $C_{12}H_{22}O_{11}$, or the more
complex starches or cellulose, represented by $(C_6H_{10}O_5)_x$. All plants,
animals and bacteria produce usable energy from stored carbon compounds by
reversing this reaction.

Photosynthesis is an extremely important and practical method of collecting
and storing solar energy and is responsible for all forms of life today.
Historically the development of man can be directly traced through
biological conversion systems, initially through the provision of food, then
food for animals, the materials for housing and energy for cooking and
heating. The commencement of industrial activities was followed by the
development of agriculture and forestry to their present levels. The renewed
emphasis on biological conversion systems arises from the fact that solar
energy can be converted directly into a storable fuel and other methods of
utilizing solar energy requires a separate energy storage system. The
carbohydrates can be reduced to very desirable fuels such as alcohol,
hydrogen or methane, a process which can also be applied directly to organic
waste materials which result from food or wood production. Compared with
other methods biological conversion efficiencies are much lower, but are
potentially far less expensive. This was recognized in many countries
during the 1970s and Hall's important review published in 1979 (21) covered
all the major research and development activities and cited nearly 150
reference sources. As with many other solar disciplines, an expansion of
the literature has taken place in parallel with these activities and a more
detailed general introduction to biological energy resources can be found,
for example, in Slesser and Lewis (22) or White and Plaskett (23).

Photosynthetic Efficiency

A comparison with most other energy conversion methods will show that
biological or photosynthetic efficiencies are very low. The efficiency is
based on the amount of fixed carbon energy produced by the plant compared
with the total incident solar radiation. Plants can only use radiation in
the visible part of the solar spectrum between wavelengths of 400-700 nm,
known as the photosynthetically active radiation (PAR) region. This
represents about 43% of the potentially available total radiation. At the
plant some of the PAR is reflected and with other losses due to internal
chemical processes the maximum attainable efficiency lies between 5 and 6%.

Under very favourable conditions, conversion efficiencies of between 2 and
5% have been recorded in the field, for example a crop of bull rush millet
in Australia achieved 4.2% over a growth period of 14 days in 1965 with
careful control of nutrient supplies. Among other peak growth rates quoted
in temperate climates are 4.3% for sugarbeet in the UK and 3.4% for maize
in Kentucky, USA. Considerably lower efficiencies are achieved over longer
periods of growth. Irish grasslands or forests with Sitka spruce are
capable of dry matter yields greater than 16 tonnes/ha which represents an
efficiency of about 0.7%. Under normal agricultural conditions very
low efficiencies are achieved, usually less than 1%. For example, the
Kentucky maize yield expressed as a function of the total annual radiation
is only 0.8%. The main reasons for these relatively low efficiencies are
environmental constraints, nutritional limitations and the incidence of
pests and diseases (15). Typical environmental constraints would include
a drought or daily variations in ambient temperature. Nutritional
limitations depend on the soil quality which, in turn, relies on the input
of fertilizers.

The importance of improving these low efficiencies was demonstrated in a
simple analysis by Hall (25). Assuming that the average solar energy
conversion efficiency could be raised to 10%, then the entire energy
consumption of the UK could be satisfied by only some 9% of the total land
area.

Energy Resources through Photosynthesis

There are three routes which can be followed to obtain the organic material
or biomass which is the starting point for the energy conversion process.
The first, and by far the simplest, is to harvest the natural vegetation.
The second, which is rapidly becoming a major 'growth' industry, is through
the cultivation of a specific energy crop, grown only for its energy content.
The third uses the wastes or by-products from plants or animals. Plants are
natural storage systems from which high grade energy can be recovered at any
time. There are fertile regions in many parts of the world where the
topography or some other reason makes the land unsuitable for agriculture or
other valuable activities. With the harvesting of natural vegetation, no
costs are involved in planting or clearing and the land would be given a new
use. A major disadvantage of this method is that the yields are at best
about half those which could be obtained from an energy plantation.

The Energy Plantation

An energy crop is grown so that its stored chemical energy can be converted
into useful energy by combustion or converted into a storable fuel. A land
crop should have as high a conversion efficiency as possible, but it does not
have to be digestible by animals or edible by humans. The entire material
or biomas of the crop can be used, including the leaves, stalks and roots.
By careful genetic selection and intensive cultivation the conversion
efficiency should reach 3% under normal conditions. An interesting
development in the UK has been the introduction of Spartina townsendii, a
type of rough grass, into intertidal mudflats. Maximum conversion
efficiencies about 50% higher than other species have been reported (26).

The use of trees as energy crops has been proposed in several countries
including the USA (27, 28), Ireland (24, 29) and Australia (30) since the
early 1970s. Detailed feasibility studies in the USA have shown that
biofuels can be produced at competitive costs, by choosing the appropriate
plant species, planting density and harvest schedule for each plantation
site, thus minimizing the overall cost of the plant material. In Ireland
about 6% of the land area consists of bogland and less than a fifth of this
area is being harvested for peat, which is either used directly as fuel in
the home or for generating electricity. During the 1970s peat provided
about a quarter of the total electrical generation. Until recently it had
been thought that bogland was unproductive, but grass, shrubs and trees have
all been successfully grown. Even with a conversion efficiency of 0.5% for
Sikta spruce, the same bogland area at present used for turf could produce
exactly half the quantity of electricity through the combustion of the trees.
An extension of the tree crop to an area about double the Irish bogland
would be sufficient to meet the entire Irish electricity demand with a
continuously renewable fuel provided that the demand did not increase. A
more detailed analysis (31) assumed that the demand for electricity nearly
doubled by the end of the century. With peat gradually being replaced by
biomass, the projected figures for electricity generation showed the
contribution from peat to be 19%, biomass 17% and other renewables (wind,
wave and photovoltaic) 17%.

An important factor in considering energy crop conversion is the energy
needed for harvesting and for fertilisers to increase the crop yields. Net
energy analysis is used to assess the energy cost/benefit ratio of any
proposed fuel conversion process. The energy inputs and outputs of the
system can be measured and the net energy ratio (NER) can be defined as the
ratio of the energy outputs to the energy inputs. Any application of this
concept requires a careful definition of the system boundaries. A detailed
analysis for an isolated community of 6000 people in Australia (22) is
significant as it draws only upon existing technology and there is no
pollution because the CO_2 fixed in photosynthesis is released in the
atmosphere after combustion. Two different systems were investigated, the
direct burning of chipped wood in a total energy steam raising system with a
steam turbine prime mover and the production of producer gas from chipped
wood with the burning of the gas in a total energy system with a
reciprocating gas engine. Both systems were considered to be technically
feasible and the major portion of the minerals required for forest regrowth
could be recycled by returning combustion ash to the forest area. An area
of less than 6 km^2 was needed, which included the additional area for the
energy required for harvesting. The case for trees as an energy crop was
put very succinctly by T. B. Reed (32) when he stated: "I would rather go
for a stroll in an acre of forest than in an acre of solar cells." With a
suitable climate it is also possible to use solar energy to dry the energy
crop. For example, a solar air heated timber kiln has operated in
Australia (33) with installations in Griffith and Townsville. Both
installations have prefabricated insulated kilns over a rock storage system.
The general conclusions were that solar drying took about twice as long as
conventional steam-heated kilns, but only half the time needed for air
drying.

The NER concept has been used in the world's first cassava (mandioca) fuel
alcohol commercial plant in Brazil (34). The system boundaries and energy
flows are shown in Fig. 8.3. The system consists of the cassava plantation,
the fuel alcohol distillery and the forest from which the fuelwood is
obtained to provide process steam for the distillery. Energy optimization
of cassava distilleries could lead to the development of varieties of
cassava with larger stalk-to-root ratios, so that the cassava stalks could
replace the fuelwood requirement. Another self-sufficient process is the
sugar cane fuel alcohol system. The bagasse or by-product can generate all
the necessary process steam. The NER of both systems is shown in Table 8.2,
taken from Ref. (34) and based on 1 m^3 of anhydrous ethanol and total on-site
generation of electric power.

Table 8.2

Raw Material	Energy (10^6 kcal)					NER
	Output	Input				
		Agriculture	Distillery	Transport	Total	
Sugar cane	5.59	0.42	0.017	0.26	0.70	8.0
Cassava	5.59	0.30	0.045	0.27	0.62	9.0

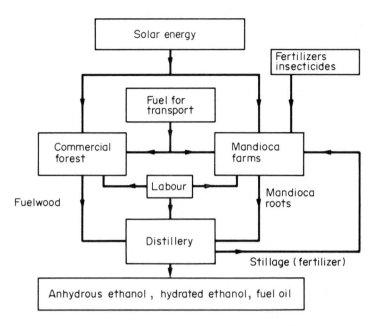

Fig. 8.3. Mendioca farms – forest – distillery.
System boundaries and energy plows.

Marine and Aqueous Applications

In oceans the production of organic matter by photosynthesis is generally
limited by the availability of nutrients and they have been compared to
deserts because of their low productivity. However, there are a few areas
where natural flows bring the nutrients from the bottom of the ocean to the
surface so that photosynthesis can take place. Particular interest has been
shown in the cultivation of giant kelp (*Macrocystis pyrifera*), a large brown
seaweed found off the west coast of the USA. An early estimate examined the
yield from an area of 600 000 km^2 and concluded that the equivalent of some
2% of the US energy supply could be provided (35). One of the disadvantages
of harvesting natural kelp beds would be the relatively low output caused by
the lack of nutrients. Artificial kelp 'farms' have been suggested (23) and
a 1000 m^2 system, with nutrient-rich deep water being pumped from deep water
to the surface kelp, has been developed, as the first stage of a project
which could lead to a 40 000 ha system.

Aquatic weeds can easily be converted to biogas. Biogas is a flammable gas
produced by microbes when organic materials, such as aquatic weeds or the
organic wastes in sewage systems, are fermented anaerobically within a
particular range of temperatures, moisture contents and acidities. It can
be used as a high quality fuel for cooking, lighting and power production.
In chemical composition it is a mixture of some 60–70% methane with carbon
dioxide and traces of other gases. The water hyacinth (*Eichhornia crassipes*)
has been extensively studied as a biogas source, particularly at the United
States National Aeronautics and Space Administration (NASA) (36). On a dry
weight basis, one kilogramme of water hyacinth can produce 0.4 m^3 of biogas
with a calorific value of 22 MJ/m^3. Aquatic weeds grow prolifically in

many tropical regions and are costly to harvest. However, as weed clearance
is essential to keep waterways clear, biogas production could be regarded
as a valuable by-product in these applications. A useful discussion of the
legal and environmental obstacles to the large-scale aquatic energy farm was
given by Wise (37).

Greenhouse Applications

Full use is made of the available solar energy in a greenhouse design
specially developed for colder regions at the Brace Research Institute (38).
It is oriented on an east-west axis with a large transparent south-facing
roof. The rear, inclined north-facing wall is insulated with a reflective
cover on the inner face. Heating requirements were reduced by up to 40%
compared with a standard, double layered plastic covered greenhouse and
increased yields of tomato and lettuce crops were reported.

An ingenious method of passive energy collection and storage for greenhouses
using the thermosyphon principle has been investigated in Australia (39).
During the day energy can be extracted from the air in a greenhouse
atmosphere and transferred to the shallow soil by passing the air through a
series of vertical pipes. The lower soil temperature will produce sensible
heat transfer and some moisture will condense and pass into the soil. The
passive system consists of a series of vertical holes in the ground with a
concentric, smaller diameter pipe, rising to a height of about 1.5 m above
ground level. In daytime, this pipe is heated by solar radiation and
produces an updraft, which in turn draws the moist air into the hole. The
method was considered to be particularly worthwhile in dry areas with a
wide range of ambient temperature from day to night and a shortage of water.

Both active and passive systems for heating the soil in greenhouses have
been studied for many years in the Soviet Union and this work has been
extensively reported in their literature. The following three examples
represent a small proportion of their current programmes. Yakubov et $al.$
(40) have suggested that at a latitude of 40°, between October and March
the maximum radiation occurs when the collector is orientated to the south
and the angle to the horizon is 52°. Their shed-type greenhouse has a south
glazed wall inclined at 52° and a long semitransparent north wall at 27° to
the horizontal. This reduces heat loses and improves shading for the sowing
area of the greenhouse. Solar energy is stored in the soil by a forced
circulation system which sends the hot air through channels some 300 mm deep.
At night the stored heat is gently released. The microclimate of a solar
greenhouse with subsoil heating has been described (41) for a greenhouse
with an enclosure coefficient of 1.53, defined as ratio of enclosure surface
to floor area. The soil heating pipes were 150 mm in diameter and laid at a
depth of 500 mm, 800-1000 mm apart. Studies were also carried out in a
$50-30^o$ greenhouse using forced electrically heated air circulation through
the channels at night and on cloudy days. In this greenhouse on a typical
January day at latitude 38° 50', an overall collection efficiency of 45% was
recorded, based on the soil storage temperature rise (42).

Some Applications in Other Fields

The photobiological production of hydrogen from water is being studied at
several centres, particularly by Hall and his colleagues at King's College,
London (43). In their work they use chloroplast membranes, hydrogenase
enzymes and other biological or synthetic catalysts.

A machine for converting grass and leaves into edible protein has been developed at the Rothamsted Agricultural Research Station (44). It can convert one ton of leaf into enough protein to meet the daily needs of 300 people by separating the fibre from the protein in leaves and grass. The edible protein has six times the protein content of an equivalent amount of beef steak. Another British development is the nutrient film technique (45) in which plants are grown in plastic troughs, closed at the top except where the plants emerge. The bottom of the troughs contains a thin film of water containing the nutrients. The method shows considerable possibilities, for the use of flat plate collectors for heating and wind energy systems for both heating and pumping. Among its advantages are that it eliminates the need for soil cultivation and sterilisation in greenhouses.

Another area in photobiological conversion research is the utilization of natural products from photsynthetic marine micro-organisms (46). Many of these organisms are capable of hydrogen photoproduction and nitrogen fixation, but the conversion efficiency is still too low. Several methods of increasing the efficiency have been suggested but a considerable amount of basic research is still needed.

In China, where approximately seven million biogas systems have been reported (47), biogas production has become a comprehensive controlled methof of waste disposal, supplying fertilizer and improving rural health in addition to providing a renewable energy source. Many of the developing countries are basing their designs on the established Chinese and Indian systems and details are regularly published. For example, an article from Bangkok in December 1981 (48) mentioned, that over 1000 gobar (cow-dung) plants have been built in Nepal and that biogas had been used to replace diesel fuel in Botswana. Plans for bringing biogas systems to over 300 000 rural farmers in Thailand are being considered in Thailand.

In the UK, the temperate climate is less encouraging for biogas, but careful overall system design can overcome this problem. A prototype unit for a dairy herd of 320 cows was completed in Kent in 1979. Electrical power was generated from a Ford diesel generator modified for gas combustion with spark ignition, with a continuous maximum power output calculated to be 25 kW (49). At least 10 anaerobic digesters were reported in the UK by the end of 1980 (50).

Conversion of Biomass to Fuels and Other Products

A selection from some of the main conversion processes is illustrated in Fig. 8.4 which shows that there are often several different routes to the same end product. A number of the processes are well known and are ideally suitable for producing fuels. With aerobic fermentation, materials containing starches and simple sugars can be used to produce ethyl alcohol or ethanol. Anaerobic fermentation has the added advantage of producing a valuable by-product, the nutrient-rich fertiliser from the digested slurry, when used to treat domestic sewage or animal wastes and produce biogas. In the pyrolysis process the organic material is heated to temperatures between 500 and 900°C at ordinary pressures in the absence of oxygen, producing methanol, which was a byproduct of charcoal in the last century. Methanol was first used as a fuel for high performance racing cars and was subsequen subsequently studied as an additive in many laboratories such as the MIT Energy Laboratory (32). It is now considered to be an essential part of the future automobile fuel mixture (51).

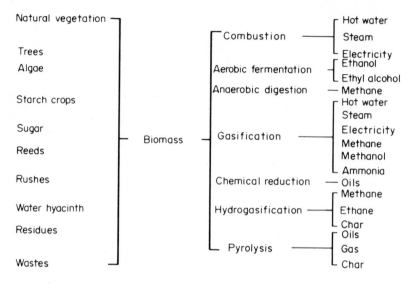

Fig. 8.4.   Some biomass sources and conversion processes.

Choice of System

The factors affecting the choice of a particular biological conversion
system identified by Hall and Coombs (52) include considerations of
agricultural capacity, environmental factors, population density, labour
intensity in the agricultural sector and the energy demand per capita. Four
regimes were distinguished in their simple classification:

(a) Temperate industrial areas such as North America, Western Europe and
     Japan, where biomass will only produce a small fraction of the energy
     demand. Emphasis will be placed on the production of scarce chemicals
     from biomass. The by-products from certain industries may be used to
     provide heat and power, while the use of wood as a direct fuel source
     is also possible.
(b) Tropical and sub-tropical regions with good soil and high rainfall such
     as parts of India and Africa, Brazil, Indo-China and north Australia.
     Energy from biomass has the greatest potential in these regions, with
     many examples already competing economically, e.g. biogas and the more
     efficient use of fast growing wood species.
(c) Northern polar and arid regions where biological systems are only
     possible in an artificial environment, e.g. use of the nutrient film
     technique.
(d) Marine and aqueous regions through the use of fast growing water weeds,
     seaweeds or micro-algae..

The development of photobiological energy conversion systems can take place more readily in the temperate western countries with their high technological background. However, these systems can function more effectively in the developing tropical and sub-tropical countries and could make a very significant contribution towards reducing their dependence on increasingly scarce and expensive oil.

PHOTOCHEMISTRY

The direct conversion of solar energy into stored chemical free energy has attracted research workers for many years. A review of this early work has been published by Archer (53) who also defined the fundamental photochemical work that remains to be done (54). Approximately half the total solar radiation which reaches the earth arrives as visible light and can be used in various photochemical reactions. The other half, which occurs in the infrared region, cannot make a useful contribution as its energy concentration is too low. The maximum overall efficiency of any photochemical energy conversion is limited to about 30%, however, as a proportion of the higher energy photons of shorter wavelengths have some energy degraded as heat during the reaction. The majority of photochemical reactions are exothermic – giving out energy – and are not suitable for converting solar radiation into stored chemical energy. The known endothermic – energy storing – reactions which occur with visible light are, in theory, capable of producing valuable chemical fuels, but a major problem has been that most of these endothermic reactions reverse too quickly to store the energy of the absorbed light. Other problems include undesirable side reactions and the high cost of the relatively scarce original material. This is relatively unimportant as the original material would be regenerated when the reaction is reversed and the stored energy is released.

One particular process which has attracted attention for many years is the possible combination of carbon dioxide and water to produce various hydrocarbons such as methane. Another possibility is the photosensitized decomposition of water to hydrogen and oxygen. This has already been achieved, although with very low efficiencies, by metal cations such as cerium and europium and the use of titanium dioxide electrodes has also been reported (55). Certain organic substances can be photoreduced in water – again with very low efficiencies.

The concept of combining the photochemical and electrochemical processes in an electrical storage battery which can be recharged directly by solar radiation is very attractive. Several examples, such as the iron-thionine system (56) are well known but have efficiencies in the order of 0.1%. In this process the bulk of the solution undergoes a photochemical change and the subsequent change in the oxidation-reduction system causes the potential. An alternative approach is to coat the electrode of the half-cell with a layer of dye or an inorganic substance such as titanium dioxide. The direction of electron flow is reversed when the electrodes are irradiated.

In an analysis of the possibilities for the photochemical conversion and storage of solar energy, Bolton (57) showed that a realistic maximum efficiency for a photochemical solar energy storage system was 15-16%. He also pointed out that the photochemical generation of hydrogen as a fuel had several attractive advantages. These included the possibility of indefinite storage, easy transport in pipelines, the use of large-scale efficient fuel cells for the generation of electricity from hydrogen and air, all associated with an almost pollution-free resource. A major review of the photobiological production of hydrogen has been given by Weaver, Lein and Seibert (58) in an important paper containing over 400 references. Discussing the possibilities of the photochemical utilization of solar energy in 1974, Porter stated that there was considerable optimism as there was such a wide range of options and a good theoretical background (55). His early optimism had not diminished by the early 1980s, as he considered that although a complete photochemical solar fuel generator had not yet been developed even on the laboratory scale, when such a reactor had been developed then scaling up to a large-scale covering many square kilometres should not involve any new problems (59). He also commented that the production of hydrogen from water had by then become considered to be 'commonplace', although problems still remained in the production of oxygen (60).

REFERENCES

(1) Telkes, M., Solar thermoelectric generators, J. Appl. Phys. 25, 765-777, 1954.

(2) Utilization of Solar Energy, Report of the NPL Committee (UK), published in *Research* 5, 522-529, 1952.

(3) *Solar energy: a UK assessment*, UK Section, ISES, London, 1976.

(4) Treble, F. C., Rapporteur's Report, Int.Conf. on Future Energy Concepts, Institution of Electrical Engineers, London, 1981.

(5) Coutts, T. J. and Hill, R., Resource constraints on the large scale terrestial exploitation of solar cells, pp. 147-150, *Ibid*.

(6) An Assessment of Solar Energy as a National Energy Resource, NSF/NASA Solar Energy Panel, 1972.

(7) Prince, M. B., Silicon solar energy converters, *J. Appl. Phys.* 26, 534-540, 1955.

(8) Wysocki, J. J. and Rappaport, P., Effect of temperature on photovoltaic solar energy conversion, *J.Appl.Phys.* 31, 571-578, 1960.

(9) Wilson, J. I. B., *Solar Energy*, Wykeham, London, 1979.

(10) Rauschenbach, H. S., *Solar Cell Array Design Handbook*, Van Nostrand Rheinhold, New York, 1980.

(11) Currin, C. C., Ling, K. S., Ralph, E. L., Smith, W. D. and Stirn, R. J., Feasibility of low cost silicon solar cells, 9th Photovoltaic Specialists Conference, Maryland, May 1972.

(12) Durand, H., Present status and prospects of photovoltaic solar energy conversion, Conf (C21), Photovoltaic Solar Energy Conversion, UK Section, International Solar Energy Society, London, 1979.

(13) Mlavsky, A. I., Press release comments, Tyco Laboratories Inc., Waltham, MA 02154, 1974.

(14) Glaser, P. E. in Arthur D. Little Inc. News, 24 October, 1981.

(15) Jesch, L. F., Solar Energy Today, UK Section, International Solar Energy Society, London, 1981.

(16) Rosenblatt, A. I., Energy crises spurs development of photovoltaic power sources, *Electronics* 47, 106, 1974.

(17) Richards, E. R., The use of solar cells in the maritime field, Proc. Conf. on Photovoltaic Cells, UK Section, ISES, November 1974.

(18) Polgar, S., Use of solar generators in Africa for broadcasting equipments, ISES Congress, Los Angeles, Extended Abstracts, Paper 13/1, July 1975, and in *Solar Energy* 19, 201-204, 1977.

(19) Haas, G. M., Bloom, S. and Cherdak, A., Experience to date with the Mitre Terrestrial Photovoltaic Energy System, Paper 21/3, *Ibid*.

(20) A Total Photovoltaic Energy System, Mississippi Community College, Blytheville, Arkansas, 1981.

(21) Hall, D. O., Solar Energy use through Biology - past, present and future, *Solar Energy* 22, 307-328, 1979.

(22) Slesser, M. and Lewis, C., *Biological Energy Resources*, E & F Spon, London, 1979.

(23) White, L. P. and Plaskett, L. G., *Biomass as Fuel*, Academic Press, London, 1981.

(24) Lalor, E., Solar Energy for Ireland, National Science Council, Dublin, February 1975.

(25) Hall, D. O., Photobiological energy conversion, *Sun at Work in Britain* 1, 14-17, 1974.

(26) Long, S., The photosynthetic potential of C₄ - plants in cool temperate ecosystems with particular reference to Spartina townsendii, Proc. Conf. Solar Energy: Biological conversion systems, UK Section, ISES and British Photobiological Society, June 1975.

(27) Szego, G. C. and Kemp, C. C., Energy Forests and Fuel Plantations, Chemtech. pp. 275-284, May 1973.

(28) Szego, G. C., Fraser, M. D. and Henry, J.-F., Design, operation and economics of the energy plantation as an alternative source of fuels, Proc. Second International Solar Forum, Vol. 2, 1-15, Deutsche Gesellschaft fur Sonnenenergie, Munich, 1978.

(29) Neenan, M., Lyons, G. and O'Brien, T. C., Short rotation forestry as a source of energy, *Solar World Forum*, Vol. 2, 1258-1262, Pergamon Press, Oxford, 1982.

(30) Burrow, A. C. and Taylor, L. E., Growing Kilowatts - bring back the axes, Ballarat Institute of Advanced Education, Australia, 1974.

(31) Towards Energy Independence, Solar Energy Society of Ireland, December 1978.

(32)   Reed, T. B., Bioconversion of solar energy, Testimony before the US
       House of Representatives Subcommittee on Energy, 18 June 1974.

(33)   Read, W. R. and Czech, J., Operating experience with a solar timber
       kiln, ISES Congress, Los Angeles, Extended Abstracts, Paper 46/2,
       July 1975.

(34)   Centro de Tecnologia Promon Newsletter, Vol. 3, No. 1, Rio de Janeiro,
       February 1978.

(35)   Wolf, M., Utilization of solar energy by bioconversion – an overview,
       Testimony before the US House of Representatives Science and
       Astronautics Committee, 13 June 1974.

(36)   Making Aquatic Weeds Useful: Some perspectives for developing countries.
       National Academy of Sciences, Washington, 1976.

(37)   Wise, D. L., Probing the feasibility of large scale aquatic biomass
       energy farms, *Solar Energy* 26, 455-457, 1981.

(38)   Lawand, T. A., Alward, R., Saulnier, B. and Brunet, E., The development
       and testing of an environmentally designed greenhouse for colder
       regions, *Solar Energy* 17, 307-312, 1975.

(39)   Morrison, G. L., Passive energy storage in greenhouses, *Solar Energy*
       25, 365-372, 1980.

(40)   Yakubov, Yu. N., Shodiev, O. Kh. and Imomkulov, A., A shed-type
       hothouse with solar energy accumulated by soil, *Geliotekhnika* 15,
       50-53, 1979.

(41)   Vardiyashvili, A. B. and Khairitdinov, B., Formation of microclimate
       in a solar greenhouse, *Geliotekhnika,* 15, 87-91, 1979.

(42)   Vardiyashvili, A. B. and Khairitdinov, B., Heat exchange in a solar
       greenhouse with a soil heating system, *Geliotekhnika,* 15, 3, 68-72,
       1979.

(43)   Gisby, P. E., Rao, K. K. and Hall, D. O., Hydrogen Production from
       water, using chloroplasts, enzymes and synthetic catalysts: a
       photobiological process, Solar World Forum, Vol. 3, 2242-2247,
       Pergamon Press, Oxford, 1982.

(44)   Making curry and haggis from leaves, *The Times,* 1 April 1976.

(45)   Cooper, A., Papers on the nutrient film technique, The Glasshouse
       Crops Research Institute, Littlehampton, Sussex, 1975-6.

(46)   Schneider, T. R., Substitute Natural Gas from Organic Materials, ASME
       Winter Meeting, New York, 27-30 November 1972.

(47)   van Buren, A. Ed., *A Chinese Biogas Manual,* Intermediate Technology
       Publications, London, 1979.

(48)   RERIC news, Renewable Energy Resources Information Center, P.O. Box
       2754, Bangkok, Thailand, Vol. 4, No. 3, December 1981.

(49) Keable, J. and Dodson, C. A. Modular System for Biogas Production Using Farm Waste, Sun 11, Vol. 1, pp. 83-87, Pergamon Press, Oxford, 1979.

(50) White, D. J., Energy for Agriculture and Forestry, in Energy in the '90s, 71-76, Highlands and Islands Development Board with the Royal Society of Edinburgh, 1980.

(51) Ward, R. F., Alcohols as fuels - the global picture, *Solar Energy* 26, 169-173, 1981.

(52) Hall, D. O. and Coombs, J., The prospect of a biological-photochemical approach for the utilization of solar energy, pp. 2-14 in Energy from the biomass, The Watt Committee on Energy, Report No. 5, London, 1979.

(53) Archer, M. D., Photochemical Aspects of Solar Energy Conversion, in *Photochemistry* 6, ed. D. Bryce-Smith, Chemical Society, Specialist Periodical Report, London, 1975.

(54) Archer, M. D., The outlook for photochemical energy conversion, ISES Congress, Los Angeles, Extended Abstracts, Paper 22/3, July 1975.

(55) Porter, G., Photochemical energy conversion, *Sun at Work in Britain* 1, 12-13, 1974.

(56) Clark, W. D. K. and Echert, J. A., Photogalvanic Cells, *Solar Energy* 17, 147-150, 1975.

(57) Bolton, J. R., Photochemical storage of solar energy, *Solar Energy* 20, 181-183, 1978.

(58) Weaver, P. F., Lein, S. and Seibert, M., Photobiological production of hydrogen, *Solar Energy* 24, 3-45, 1980.

(59) Porter, G., Prospects for a biological synthesis of "biomass" Paper V/Kl Proc.Conf. 'Energy from Biomass', Brighton 1980, Applied Science Publishers, 1981.

(60) Porter, G., Photochemical Routes to Solar Energy Retrieval, *Solar World Forum*, Vol. 3, 2181-2185, Pergamon Press, Oxford, 1982.

CHAPTER 9

WIND POWER

INTRODUCTION

Energy from the wind is derived from solar energy, as a small proportion of
the total solar radiation reaching the earth causes movement in the
atmosphere which appears as wind on the earth's surface. The wind has been
used as a source of power for thousands of years, both on land and at sea.
Sailing ships were first reported in ancient Egypt nearly five thousand years
ago and reached their peak towards the middle of the nineteenth century with
the development of the fast international trading clipper ships. However, by
the beginning of the twentieth century fossil-fueled steam-engined ships had
become firmly established and although the wooden sailing ships of that era
could compete with steam, they continued to decline as the engine-powered
steel ships improved technologically and by the 1930s only a few large
sailing ships remained.

Windmills for mechanical power on land may have first appeared in Persia,
where archaeologists have found evidence of the use of wind-driven water
pumps for irrigation dating from about the fifth century. These early
Persian designs used cloth sails and had a vertical axis, the vertical sails
on one side caught the wind while it spilled out on the other side. With
the vertical axis there is no difficulty in steering the sails or blades to
face the wind. The traditional horizontal axis tower mill for grinding corn,
with sails supported by a large tower rather than a single post, had been
developed by the beginning of the fourteenth century in several parts of
Europe. Its use continued to expand until the middle of the nineteenth
century when the spread of the steam engine as an alternative, cheaper,
source of power led to its decline. Rural areas in the United States
experienced a similar situation at the beginning of the twentieth century.
Thousands of farms had steel-towered windmills to pump water and in some
cases generate electricity at that time, but over the next fifty years the
rural electrification scheme became established and the great majority of
these early windmill systems were allowed to fall into disuse. The success
of these devices is illustrated by the estimated figures of 50 000 wind/
electric generators, or aerogenerators, supplied over the period (1, 2).

HISTORICAL DEVELOPMENT OF WIND-GENERATED ELECTRICITY

Denmark

Towards the end of the last century, the windmill was the principal source
of power in agricultural areas of Denmark. Known as house-mills, they were
often mounted on the roofs of barns and, together with industrial mills,
were estimated to be producing about 200 MW (3) from over 30 000 units. In
about 1890, Professor P. La Cour commenced work on wind-power and obtained
substantial support from the Danish Government, which not only enabled him
to erect a windmill at Ashov, but provided a fully instrumented wind tunnel
and laboratory. Between 1890 and his death in 1908, Professor La Cour
developed a more efficient, faster running windwheel, incorporating a
simplified means of speed control, and pioneered the generation of
electricity. The Ashov windmill had four blades 22.85 m (75 ft) in diameter,
on a steel tower 24.38 m (80 ft) high. Power was transmitted, through a
bevel gearing, to a vertical shaft which extended to a further set of bevels
at ground level, and the drive was connected to two 9 kW d.c. generators –
the first recorded instance of wind-generated electricity. By 1910 several
hundred windmills of up to 25 kW capacity were supplying villages with
electricity.

The use of wind-generated electricity continued to increase and during the
1939-1944 war period, a peak of 481 785 kWh was obtained from 88 windmills
in January 1944 (4). Included among these 88 was the F. L. Smidth unit
erected in Gedser in 1942. Originally it was a 70 kW d.c. unit, with three 24.38 m
(80 ft) diameter wooden blades, it was converted to a.c. in 1955. It
produced approximately 700 000 kWh during its first five years in operation,
or about 2 000 kWh per annum/rated kW.

A detailed historical survey of the number of wind machines in Denmark from
1900 to 1950 has been given by Ølgaard (5).

The United States

By 1922 the Farm Light and Power Year Book listed 54 different manufacturers
of windmill pumps and electrical plant. Towards the end of that decade one
of the most successful windmill manufacturing companies was established –
the Jacobs Wind Electric Company of Minneapolis, Minnesota (6). The company
was founded by M. L. Jacobs who introduced two significant innovations in
his designs; (a) a three-bladed propellor which effectively eliminated
vibrations found in two-bladed systems due to variations in the total forces
acting on the blades as they moved from the horizontal to the vertical
position, and (b) a flyball governor to control the pitch of the blades,
allowing them to 'feather' when the wind velocity was greater than 8.05 m/s
(18 mph) and maintain a constant speed to drive the generator. The blades
had a diameter of about 4.27 m (14 ft) and were directly coupled to the
generator without gearing. Perhaps the best-known application was in the
Antarctic, where Admiral Byrd installed a Jacobs generator during one of his
scientific expeditions in the 1930s. The unit was still functioning when
Byrd returned in 1946. The rural electrification scheme forced the company
out of business in 1957.

The Smith-Putnam Windmill (7)

Conceived by an American engineer, Palmer C. Putnam, in the 1930s, the Smith-Putnam Windmill had two blades with a diameter of 53.34 m (175 ft) and was erected at Grandpa's Knob in central Vermont in 1941. At that time it was the world's largest ever windmill and was to hold this record for the next 35 years. The synchronous electric generator and rotor blades were mounted on a 33.54 m (110 ft) tower and electricity was fed directly into the Central Vermont Public Service Corporation network. The windmill was rated at 1.25 MW and worked well for about 18 months until a main bearing failed in the generator, a failure unconnected with the basic windmill design. It proved impossible to replace the bearing for over two years because of the war and during this period the blades were fixed in position and exposed to the full force of the wind. During the original assembly of the mainly stainless steel blades and supporting spars, rivet holes had been drilled and punched in the blades and cracks had been noticed in the metal around the punched holes in 1942. It was decided to carry out repairs on site, rather than returning the whole assembly to the factory. On 26 March 1945, less than a month after the bearing had been replaced, the cracks widened suddenly and a spar failed causing one of the blades to fly off. The S. Morgan Smith Company, who had undertaken the project, decided that they could not justify any further expenditure on it, apart from a feasibility study on the installation of other units in Vermont. This indicated that the capital cost per installed kilowatt would be some 60% greater than conventional systems.

Although sceptics have tended to regard this experiment as an expensive failure, it was the most significant advance in the history of windpower. For the first time synchronous generation of electricity had taken place and been delivered to a transmission grid. Both mechanical failures were due to a lack of knowledge of the mechanical properties of the materials at that time. Bearing design and the problems of fatigue in metals have been studied extensively since then and similar failures are unlikely to occur in modern windmills. Their research programme included an extensive series of on-site measurements, which proved that the actual site at Grandpa's Knob had a mean wind velocity of only 70% of the original estimated velocity* and that many other sites should have been selected. The technical problems of converting wind energy into electricity had been largely overcome and the possibility of developing wind power as a national energy resource in any country with an appropriate wind climate had been established.

Russia

In 1931, the Russians built the first windmill to feed electricity directly into an a.c. network at Yalta, near the Black Sea (8). Used as a supplementary power source, it was connected to a conventional fossil-fuel plant at Sevastopol, about 30 km away. It had three blades 30.48 m (100 ft) in diameter driving a 100 kW induction generator through wooden gears. The tower was 30.48 m (100 ft) high but was provided with an inclined strut to carry the thrust of the wind from the top of the tower to the ground. The base of the strut was driven round a circular track by an electric motor controlled by a wind direction sensing vane at the top of the tower. The metal-covered blades could be feathered by an automatic pitch control system

---

*The measured annual output of about 30% of the predicted performance confirmed the cube law relationship as $0.7^3 = 0.343$.

activated by the effect of centrifugal force on offset flaps, so that the
plant could continue to operate in high winds at an approximately constant
speed. An annual output of 279 000 kWh was reported from the site which had
an annual mean wind velocity of 6.7 m/s (15 mph), but satisfactory control
was difficult to achieve. Over the next two decades developments in Russia
were limited to plants generating up to 3 kW (6).

The United Kingdom

By the 1920s, interest in small wind powered electrical generators had been
well established. Comparative test results on seven different commercially
available windmills ranging in power from 250 W to 10 kW had been published
(9) and also a handbook for practical engineers who wished to build their
own machines (10). During the 1930s, the Lucas 'Freelite' was developed
(11), intended for use with up to six lighting points - three 40 W and three
25 W bulbs at 25 volts. On the Freelite, the rotor could be turned out of
high winds by means of a furling handle at the base of the tower.

Two 100 kW machines were built shortly after the war. The first, in 1950,
was built by John Brown and Company and erected in the Orkneys (12), had 3
15.24 m (50 ft) diameter blades mounted on a 23.77 m (78 ft) tower driving
a 100 kW a.c. induction generator. The second was built for Enfield Cables
by deHavilland Propellors and the Redheugh Iron and Steel Company and
featured a pneumatic transmission system invented by Andreau in France (13).
The two 24.38 m (80 ft) diameter blades were hollow with a hole at the tip
so that during rotation they acted as a centrifugal pump. The induced
internal air flow entered the base of the tower through a turbine, directly
coupled to a synchronous generator. Originally erected at St Albans in
1953, it could not be tested there because of the poor wind climate, and was
subsequently re-erected in Algeria for Electricité et Gaz d'Algérie in 1957.
Its full rated output of 100 kW was obtained shortly after re-erection.
Studies were also carried out on the evaluation of windmill performance and
a graphical method for predicting the performance of wind powered electrical
generators was established by the Electrical Research Association (14) in
1960, based on results obtained from two windmills, a three bladed 12.19 m
(40 ft) diameter unit rated at 25 kW erected on the Isle of Man and a three
bladed 10 m diameter unit rated at 7.5 kW installed in Scotland. Excellent
agreement between the experimental results and the predicted performance was
obtained in each case.

WIND ENERGY POTENTIAL

Wind has a dependable annual statistical energy distribution but a complete
analysis of how much energy is available from the wind in any particular
location is rather complicated. It depends, for example, on the shape of
the local landscape, the height of the windmill above ground level and the
climatic cycle. Somewhat surprisingly, the British Isles have been studied
more extensively than practically any other country in the world (11, 15)
and the west coast of Ireland, together with some of the western islands of
Scotland, have the best wind conditions with mean average wind speeds
approaching 9 m/s.

The kinetic energy of a moving air stream per unit mass is $\frac{1}{2}V^2$ and the mass flow rate through a given cross-sectional area A is $\rho A V$. The theoretical power available in the air stream is the product of these two terms

$$\frac{1}{2}\rho A V^3.$$

If the area A is circular, typically traced by rotor blades of diameter D, then $\frac{\pi}{4}\ D^2 = A$, and the power available becomes

$$\frac{\pi}{8}\rho D^2 V^3.$$

The actual power available can be conveniently expressed (15) as

$$K_r D^2 V^3$$

where K_r is a term specifically associated with wind dynamics and the efficiency of the rotor power system.

The maximum amount of energy which could be extracted from a moving airstream was first shown by the German engineer Betz, in 1927, to be 0.59259 of the theoretical available power. This efficiency can only be approached by careful blade design, with blade-tip speeds a factor of six times the wind velocity. Any aerogenerator will only operate between a certain minimum wind velocity, the starting velocity V_s, and its rated velocity V_R. Typically V_R/V_s lies between 2 and 3. If the pitch of the blades can be altered at velocities greater than V_R, the system should continue to operate at its rated output, the upper limit depending only on the design. In some systems the whole rotor is turned out of the wind to avoid damage at high wind speeds. An annual velocity and power duration curve for a continuously generating windmill is shown in Fig. 9.1. Many current designs give a rotor conversion efficiency of 75%.

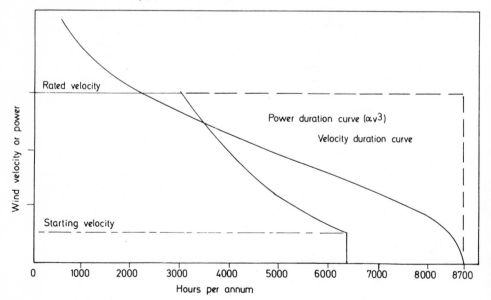

Fig. 9.1. Annual velocity and power duration curves.

Taking the air density, ρ, as 1.201 kg/m^3 at normal atmospheric pressure (1000 millibars) and 290 K and assuming a rotor conversion efficiency of 75%.

$$K_r = \frac{\pi}{8} \times \frac{1.201 \times 0.593 \times 0.75}{1000} \text{ (or approximately 0.00020).}$$

The effect of the height of the windmill tower on the performance can be significant and empirical power law indices have been established (16), relating the mean wind velocity V to the height H, in the equation $V = H^a$. A value of $a = 0.17$ is the accepted value in the UK for open, level ground, but this rises to 0.25 for an urban site and 0.33 for a city site. An ideal site is a long, gently sloping hill. Methods for determining the probable mean wind velocity and energy multiplication factors have been described by Caton (17) and Rayment (18).

Using the value for K_r derived above, the power can be converted to an annual mechanical or electrical output, E_a, as follows:

$$E_a = K_r D^2 V^3 \times K_s H \text{ (kWh)}$$

where H is the average number of hours in a year (8766) and K_s is a semiempirical factor associated with the statistical nature of wind energy recovery.

The mean annual wind velocity is normally used to describe the wind regime at any particular location, but the output from a windmill is proportional to V^3. Since a transient arithmetic increase in wind velocity will contribute much more energy to the rotor than an equal arithmetic decrease will deduct, the mean of V^3, which is always much greater than the cube of the mean annual wind velocity, should be used. A value of $K_s = 1.20$ was suggested by Juul (3) in 1956, who used a mean velocity of 8 m/s as a reference and considered that the most common variation in wind velocity occurred at frequent short intervals between 6 m/s and 10 m/s, 8^3 equalling 512, whereas $\frac{1}{2}(6^3 + 10^3) = 608$. A more recent computer analysis by Pontin (19) in 1975 suggests that a value of 2.06 could be taken, giving an approximate value for $K_r K_s$ as 0.0004 and $K_r K_s H$ becomes 3.5064. This value is very close to the figure derived by Rayment, based on data published in Met 0792 (20) and Caton's analysis (17), where the annual extractable energy, if the rotor shaft is connected to an electrical generator, is given by

$$E_a = 0.0148 \ V_{50}^3 \text{ (GJ/m}^2\text{)}$$

or

$$3.2289 \ D^2 V_{50}^3 \text{ (kWh).}$$

V_{50} is the velocity exceeded for 50% of the year and is quite close to the mean annual wind speed. Using the above equation, Table 9.1 shows the total power which could be produced per annum from a windmill with an 18.3 m diameter rotor.

Table 9.1

Mean wind velocity (m/s)	kWh/annum	Mean wind velocity (m/s)	kWh/annum
4	75 153	9	856 034
5	146 782	10	1174 258
6	253 640	11	1562 983
7	402 771	12	2027 118
8	601 220		

SOME RECENT DEVELOPMENTS

The Vertical Axis Windmill (curved blades)

The modern vertical axis windmill is a synthesis of two earlier inventions;
(a) the Darrieus (21) windmill with blades of symmetrical aerofoil cross-
section bowed outward at their mid-point to form a catenary curve and
attached at each end to a (vertical) rotational axis perpendicular to the
wind direction and (b) the Savonius (22, 23) rotor or S-rotor, in which the
two arcs of the 'S' are separated and overlap, allowing air to flow through
the passage. Simple Savonius rotors have been made out of two standard oil
drums cut in half and welded together to form the blades (24). The Darrieus
windmill is the primary power-producing device, but, like other fixed-pitch
high-performance systems, is not self-starting. The blades rotate as a
result of the high lift from the aerofoil sections, the S-rotor being used
primarily to start the action of the Darrieus blades. The wind-energy
conversion efficiency of the Darrieus rotor is approximately the same as any
good horizontal system (25) but its potential advantages are claimed to be
lower fabrication costs and functional simplicity (26). A major investigation
of the system was carried out by the Sandia Laboratories (1), which resulted
in the development of a machine with a 17 m diameter rotor by 1977. Four
years later, in 1981, the largest American Darrieus machine, with three
blades, was developing 500 kW (27). Developments have also been reported in
Sweden (28) and Germany (29), while in Canada a 3 MW machine, developed
jointly by the government and the Hydro Quebec utility company was due for
completion in 1984.

The Vertical Axis Windmill (straight blades)

An analysis of the Darrieus rotor suggested to Musgrove (30) that straight-
bladed H-shaped rotors, with the central horizontal shaft supporting two
hinged vertical blades, could be a more effective system. A variety of
designs based on Musgrove's work in the UK during the 1970s and early 1980s
have been studied and a small, 6 m diameter, three-bladed version was
commercially available by 1980. Musgrove also considered the possibility of
siting groups or clusters of windmills in shallow off-shore locations in the
UK such as the Wash. Two advantages of this proposal are the higher mean
wind speeds and the greatly reduced environmental objections.

Another vertical axis system which was originally conceived over fifty years
ago, was re-examined in the United States and shown to be theoretically
competitive with horizontal axis machines of a comparable size (31). This
is the Madaras Rotor Power Plant, which utilizes rotating cylinders,

vertically mounted on flat cars placed on a closed track, to react with the
wind and drive the train of cars round the track. A rotating cylinder
generates a lift force about ten times larger than a conventional aerofoil
or windmill blade. In the original experiments, carried out between 1929
and 1934, a 27 m high cylinder, 6.8 m in diameter, with a maximum rotational
speed of 160 rev/min was tested in New Jersey. Among the advantages claimed
for the system, compared with large horizontal axis machines, are superior
durability and structural life, better gust resistance and a higher cut-out
wind speeds. A straight track, some 20 000 m long, with semi-circular turns
at each end was among 12 different layouts studied. With 190 rotors, over
200 MW could be generated.

Denmark

The first windmill of the 1970s and 1980s to hold the record as the largest
in the world was the Twind windmill, located near the west coast of Jutland
(32). A distinctive feature of the project was that it was not sponsored
by the Danish government but resulted from a team effort from a college
community who provided their own labour and finance. As shown in Fig. 9.2,
it has three rotor blades with an overall diameter of 54 m and a reinforced
concrete tower with a tower hub height of 54 m above ground level.
Originally designed with a starting velocity of 3 m/s and a rated velocity
of 14 m/s, it was rated at 2 MW. Normally its output is less than 1 MW and
this is used to heat a large insulated 3000 m^3 hot water storage tank.

The official Danish programme for large electricity producing wind energy
systems started in 1977 with a joint programme directed by the Energy
Ministry and the Electricity Utilities. The first project was the
refurbishing of their 24 m diameter Gedser machine, which was originally
commissioned in 1957. This completed a further two year test programme in
1979. The major project was the design, construction and operation of two
much larger machines, known as Nibe A and Nibe B, which were erected in 1979.
These turbines are sited close to each other and are identical, apart from
their rotor blades. Those for the A machine are supported by stays while
the blades of the B machine are self-supporting. The main features are as
follows:

  Three-blades, upwind rotor, 40 m diameter.
  The A machine has 4 discrete pitch angles (start, stop, low wind operation,
  high wind operation) and is stall-regulated. The B machine has full pitch
  control for power regulation.
  Blade construction: steel/fibre glass spar, fibre glass shell.
  Four-pole asynchronous generator, rated power 630 kW.
  Concrete tower, height to rotor hub 45 m.
  Wind speeds: cut in 6.5 m/s (at hub height), rated 13-14 m/s, cut-out
  25 m/s.

By the end of 1980 both machines had been connected to the grid. It is
thought that it would be possible to site between 1000 and 2000 wind turbines
on land in Denmark, with an annual production equivalent to about 10% of the
1980 Danish electricity consumption (33).

The problems associated with power fluctuations in wind energy systems in
Denmark were analysed by Sorensen (34) who concluded that the addition of
short-term storage, capable of delivering the average power for 10-20 hr,
would make the wind energy system as dependable as one large nuclear plant,
being capable of delivering the average power for about 70% of the time.

Fig. 9.2.

Federal Republic of Germany

The German wind programme, known as the Growian programme, had some 25
projects in operation by 1980, ranging from some small, low-cost units rated
at 15 kW for production in developing countries and a medium sized 52 m
diameter twin-bladed 265 kW machine - the Voith-Hütter, commissioned in 1981
- to the giant MW range Growian I and Growian II machines.

Growian I is located at the Kaiser-Wilhelm-Koog at the mouth of the Elbe and will be the largest windmill in the world by the end of 1982 with a blade diameter of 100 m and a hub height above ground of 100 m. It will be operated by the "GROWIAN-Bau-und-Betriebs-GmbH", a company consisting of the three German utilities HEW, RWE and Schleswag. It is a twin-bladed horizontal axis machine, designed to generate 3 MW at its rated speed of 11.8 m/s. The cut-in speed is 6.3 m/s, cut-out speed 24 m/s and the designed rotor speed is 18.5 rev/min, giving an estimated 12 GWh per annum.

A second project, Growian II, was under development in 1982, with a third-scale prototype leading to a single bladed, horizontal axis machine with an overall diameter of 145 m. The main features of the Growian II project are as shown in Table 9.2.

Table 9.2

	Third-scale prototype	Growian II
Rotor disc diameter (m)	48	145
Hub height (m)	50	120
Power at rated wind speed (kW$_{el}$)	370	5000
Power density (W$_{el}$/m^2)	200	300
Rated wind speed (m/s)	10	11.5
Operational wind speed range (m/s)	5.7 – 16	6.6 – 18
Annual power production (GWh/a)	1.3	17
Blade tip speed at rated wind speed (m/s)	120	138

The Growian programme has been criticized by Hugosson (33) who felt that too much was being attempted and that Growian I would not be operational before 1983.

The Netherlands

Although the traditional Dutch windmill was well established hundreds of years ago, their current (1980s) programme is relatively modest. Starting with feasibility studies, including work on off-shore siting, their main research windmill is at the Energiecentrum Nederland (ECN), Petten. This is a horizontal axis, upwind, two-bladed unit rated at 125 kW. The siting of the unit within a busy research establishment required an elaborate risk analysis, which led to the conclusions that a catastrophic event with one victim could occur with a frequency of once in some 300 years.

One major initiative by a private organization being studied by a government commission in 1981 was a proposal to store surplus windpower and over-capacity night power in artificial lakes. These would be situated adjacent to the polders, have a typical area of 300 km^2, with a water level only 16-20 m above the polder. One such artificial lake could have an installed hydro power of 2000 MW.

Sweden

The main feature of the Swedish programme was related to the design, construction and operation of two large-scale prototypes, located at Maglarp in the province of Skane in southern Sweden, and Näsudden on the island of Gotland. These projects aim at obtaining a basis for a decision in 1985 about the future of wind power in Sweden. A smaller experimental unit at Kalkugnen WTS 1, rated at 60 kW, gave valuable data before an accident occurred during a planned change of rotor blades. Details of the large machines are shown in Table 9.3.

Table 9.3

	Maglarp (WTS 3)	Näsudden (WTS 2)
Rated power (kW)	3000	2000
Turbine diameter (m)	78	75
Hub height (m)	80	80
Turbine position	Downwind	Upwind
Tower materials	Steel	Concrete
Cut-in wind speed (m/s)	6	6
Rated wind speed (m/s)	14	13
Cut-out wind speed (m/s)	21	21
Annual output (GWh)	7-8	6-7
First rotation planned	Nov. 1981	June 1982

The United States

The first major project in the USA wind energy programme was the ERDA Model Zero (MOD-0) 100 kW windmill which consisted of a two-bladed, 38.10 m (125 ft) diameter, variable-pitch propeller system driving a synchronous alternator through a gearbox, mounted on a 30.48 m (100 ft) steel tower (2, 35, 36). The blades were located downstream from the tower and a powered gear control system replaced the traditional tail fin of earlier designs. Power generation commenced when the wind speed reached 3.58 m/s (8 mph) and reached its rated 100 kW at 8.05 m/s (18 mph), a V_R/V_S ratio of 2.25. The maximum blade rotational speed was 40 rpm and was maintained at higher wind speeds by varying the blade pitch angle.

This initial test programme was designed to establish a data base concerning the fabrication, performance, operating and economic characteristics of propeller-type wind turbine systems for providing electrical power into an existing power grid. Subsequently this machine was modified and uprated to 200 kW and re-erected at Clayton, New Mexico, where it generates about 3% of the town's electricity. Three similar machines were erected at other sites in Perto Rica, Rhode Island and Hawaii.

The MOD-1 windmill became the world's largest machine in May 1979, when it was commissioned. This was also a twin-bladed horizontal axis machine with a blade diameter of 60.96 m (200 ft) and rated at 2 MW. Minor problems of interference with televison signals and a low level, low frequency noise were easily overcome. This was followed in December 1980 with the commissioning of the first of three 2.5 MW MOD-2 machines at Goldendale, Washington. These kept to the twin-bladed horizontal axis design with a

blade diameter of 91.44 m (300 ft) and by grouping three on a fairly small
site, it was possible to study the interaction between a windmill cluster.
These were the largest windmills operating in the world during 1981. As
with earlier versions, minor problems occurred during the first year of
operation, with an overload causing damage to the shaft connected a blade
to the gearbox of one machine in 1981, resulting in modifications to all
three machines.

The United Kingdom

There was considerable interest in March 1976 when the Electrical Research
Association in England, in evidence to the House of Commons Select Committee
on Energy, stated that giant windmills could generate up to 10% of the UK
electricity requirement within ten years and quoted approximately 1500
windmills as a guideline. However, very little happened over the next few
years apart from the establishment of a few very small companies
manufacturing small (less than 25 kW) machines and two medium sized privately
financed projects - the three-bladed Aldborough 17 m diameter machine built
by Sir Henry Lawson-Tancred in 1977 and the Wind Energy Supply Company's
twin-bladed 18.3 m diameter unit, with a direct hydraulic transmission from
the top of the tower to the point of use, which was briefly tested in 1976.
The Aldborough machine was directly connected to the national grid and this
operating experience led to a larger 65 kW version in 1980, but the Wind
Energy Supply Company's initiative failed through lack of further development
funding.

A design feasibility and cost study of large wind turbine generators suitable
for network connection was carried out in 1976 and 1977 by a group comprising
of British Aerospace Dynamics Group, Cleveland Bridge and Engineering Co Ltd,
Electrical Research Association Ltd, North of Scotland Hydro-Electric Board,
South of Scotland Electricity Board and Taylor Woodrow Construction Ltd (37).
A reference design was evolved for a 60 m diameter turbine in 1977. It was
rated at 3.17 MW at 22 m/s (hub height) and had two fixed pitch steel blades
driving an induction generator. By February 1980 a two metre diameter model
had been tested in a wind tunnel. Fortunately the North of Scotland Hydro-
Electric Board suddenly realized there was an urgent need to implement wind
energy conversion systems on three island systems currently using diesel
generated electricity, Orkney, Shetland and the Western Isles. Of these,
Orkney was preferred as the location for the first UK Department of Energy
demonstration, wind data having suggested the availability of a large number
of potential sites on the island that were amongst the best in the UK. A
20 m diameter prototype, which models the main features of the 60 m machine,
generating some 250 kW, had been designed by 1981 and it was planned to
connect this to the Orkney grid in time for any necessary modifications to
be made to the 60 m machine. As the diesel system on Orkney represents a
set of operational criteria typical of many hundreds of similar systems
around the world, it was thought that this would give an opportunity to
demonstrate machines for a wide international market.

Other European countries with wind power programmes include Italy, Norway
and Spain where a 150 kW windmill was being developed in 1981. The essential
feature of all these various proposals is that they are based on existing
technology, or a technology that is clearly within grasp.

Russia

Wind power development in the Soviet Union is organized by the Tsiklon
(Cyclone) Institute. Reports of wind power developments which appear in the
Soviet press and in overseas broadcasts suggest that both small (up to
10 kW) and medium (10-100 kW) range machines are being used for a variety of
applications such as water pumping, desalination and cathodic protection.
Several large units, of up to 5 MW capacity are believed to be under
development and a proposal for a network of some 200 machines rated at 1 MW
for the remote and sparsely populated region between the White Sea and the
Barents Sea was reported in 1981.

The Wind-Solar Approach

The complementary nature of wind and solar energy in the British Isles is
illustrated in Fig. 9.3, based on approximate maps of solar radiation (38)
and windpower (18). This shows that although the annual mean values of
global radiation are lower in the north, there is a very much greater
potential in wind energy. In many countries it has been found that high
winds occur most frequently in the winter months when the demand for energy
is also at its peak. It is also the period when solar radiation is at its
lowest for direct water or space-heating applications. One solution could
be to combine the availability of wind with a solar heating system through
the use of ducted rotors. Ducted rotors which are free to rotate into the
wind can theoretically extract much more power from a given wind than an
unducted rotor of the same diameter. For example, a figure of 46% more
power for a 3.5 m diameter ducted rotor has been given by Lewis (39) and 65%
by Lilley and Rainbird (40). If these ducted rotors are not placed on a
tower, but arranged in fixed banks to form a 'wind-wall', an aesthetically
satisfying arrangement is obtained which avoids the visual obtrusion of
isolated large windmill units into an urban landscape. One such arrangement
has been proposed (41) for a housing scheme in Sussex, where long-term heat
storage would be provided by underground water storage tanks and the cut for
this storage used to provide slopes or banks on which solar collectors can
be placed. The wind-wall is placed at the top of the slope, as shown in
Fig. 9.4, with a group of garages erected over the underground storage
system. The fixed direction of the ducted rotor, together with their height,
give an estimated efficiency factor of 77% compared with a similar sized
normal windmill.

An alternative to the ducted rotor could be the use of a delta-windshape to
generate vortices, an approach to power augmentation described by Greff and
Hozdeppe (42).

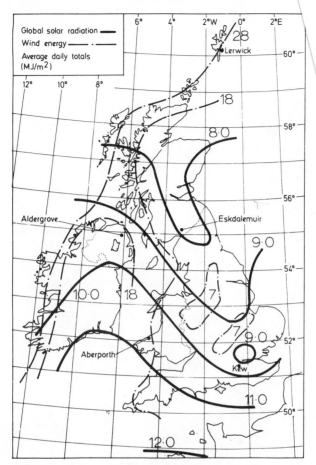

Annual mean global daily irradiation (MJ/m²) and daily wind energy
available at the rotor (MJ/m²)

Fig. 9.3.

Fig. 9.4.

REFERENCES

(1) Blackwell, B. F. and Feltz, L. V., Wind energy - a revitalized pursuit,
     SAND-75-0166, Sandia Laboratories Energy Report, March 1975.

(2) Reed, J. W., Maydew, R. C. and Blackwell, B. F., Wind energy potential
     in New Mexico, SAND-74-0077, Sandia Laboratories Energy Report, July
     1974.

(3) Juul, J., Wind Machines, Wind and Solar Energy Conference, New Delhi,
     UNESCO, 1956.

(4) Golding, E. W. and Stodhart, A. H., The use of wind power in Denmark,
     ERA Technical Report C/T 112, 1954.

(5) Ølgaard, P. L., On the number of wind machines in Denmark from 1900
     to 1950, *Solar Energy* 22, 477-478, 1979.

(6) Clark, W., *Energy for Survival*, p. 539, Anchor Press, New York, 1975.

(7) Putnam, P. C., *Power from the Wind*, Van Nostrand, New York, 1948.

(8) Gimpel, G., The windmill today, ERA Technical Report IB/T22, 1958.

(9) Cameron Brown, C. A., Windmills for the generation of electricity,
     Institute for Research in Agricultural Engineering, Oxford University,
     1933.

(10) Powell, F. E., *Windmills and wind motors*, Percival Marshall, London,
     1928.

(11) Golding, E. W., *The generation of electricity by wind power*, E. & F.
     Spon, 1955. Reprinted CTT 1976.

(12) Venters, J., The Orkney windmill and wind power in Scotland, *The
     Engineer*, 27 January 1950.

(13) Wind generated electricity prototype 100-kW plant, *Engineering* 179,
     4652, 28 March 1955.

(14) Tagg, J. R., Wind driven generators: The difference between the
     estimated output and actual energy obtained, ERA Technical Report C/T
     123, 1960.

(15) Golding, E. W. and Stodhart, A. H., The potentialities of windpower for
     electricity generation, British Electrical and Allied Industries
     Research Association, Tech. Rep. W/T16, 1949.

(16) Davenport, A. G., Proceedings of the (1963) Conference on Wind Effects
     on Building and Structure Vol. 1, HMSO, 1965.

(17) Caton, P. G., Standardised maps of hourly mean wind speed over the
     United Kingdom and some implications regarding wind speed profiles,
     Fourth International Conference on Wind Effects on Building and
     Structure, London, 1975.

(18) Rayment, R., Wind energy in the UK, *The Building Services Engineer*, 44,
     63-69, June 1976.

(19) Pontin, G. W.-W., The bland economics of windpower, Wind Energy Supply
      Company, Peacehaven, Sussex, 1975.

(20) Tables of surface wind speed and direction over the United Kingdom,
      Meteorological Office, Met 0792, HMSO, 1968.

(21) Darrieus, G. J. M., Turbine having its rotating shaft transverse to
      the flow of the current, US Patent 1,835,018, 8 December 1931.

(22) Klemin, A., The Savonius wing rotor, *Mechanical Engineering* 47, No. 11,
      November 1925.

(23) Savonius, S. J., The S-rotor and its application, *Mechanical Engineering*
      53, No. 5, May 1931.

(24) Brace Research Institute, McGill University, Montreal, Canada.

(25) South, P. and Rangi, R. S., A wind-tunnel investigation of a 14 ft
      diameter vertical-axis windmill, National Research Council of Canada,
      LTR-LA-105, September 1972.

(26) South, P. and Rangi, R. S., The performance and economics of the
      vertical-axis wind turbine developed at the National Research Council,
      Ottawa, Canada, *Agricultural Engineer*, February 1974.

(27) Jesch, L. F., Solar Energy Today, UK Section ISES, p. 157, August 1981.

(28) Ljungström, O., 'L-180 Poseidon: A new system concept in vertical axis
      wind turbine technology', Proc. 3rd International Symposium on Wind
      Energy Systems, BHRA Fluid Engineering, Cranfield, pp. 333-355, 1980.

(29) Fritzche, A. and Wirths, G., 'Wind Energy Converter with Vertical
      Rotation Axis', Proc. Conf. (M2) Solar Energy in the 80s, UK-ISES
      Midlands Branch, pp. 16-21, 1980.

(30) Musgrove, P. J. and Mays, I. D., 'The Variable Geoemtry Vertical Axis
      Windmill', Proc. 2nd Int. Symposium on wind energy systems, BHRA Fluid
      Engineering, Cranfield, E4, pp. 39-60, 1978.

(31) Whitford, D. H. and Minardi, J. E., Utility-sized Madaras wind plants,
      The International Journal of Ambient Energy, 2, pp. 3-21, 1981.

(32) Hinrichsen, D. and Cawood, P., Fresh breeze for Denmark's windmills,
      New Scientist, 567-570, 10 June 1976.

(33) Hugosson, S., The European Windpower Scene, Proc. 3rd BWEA Wind Energy
      Conference, Cranfield, 7-15, 1981.

(34) Sorensen, B., On the fluctuating power generation of large wind energy
      converters, with and without storage facilities, *Solar Energy*, 20,
      No. 4, 321-331, 1978.

(35) Kocivar, B., World's biggest windmill turns on for large-scale wind-
      power, *Popular Science*, March 1976.

(36) Puthoff, R. L. and Sirocky, P. J., Preliminary design of a 100 kW wind
      turbine generator, NASA, NASA TM X-71585, August 1974.

(37) Lindley, D. and Stevenson, W., The horizontal axis wind turbine project on Orkney, *Proc.*, Third BWEA Wind Energy Conference, Cranfield, 16-23, 1981.

(38) Solar Energy: a UK assessment, UK Section ISES, London, May 1976.

(39) Lewis, R. I., Wind power for domestic energy, Appropriate Technology for the UK, University of Newcastle upon Tyne, March 1976.

(40) Lilley, G. M. and Rainbird, W. J., A preliminary report on the design and performance of directed windmills, ERA Technical Report C/T 119, 1957.

(41) McVeigh, J. C. and Pontin, G. W. W., The wind-wall - an integrated wind solar system, *Wind Engineering*, 1, 150, 1977.

(42) Greff, E. and Holzdeppe, D., 'Wind Energy Concentration with Vortex Flow Fields, and its Utilisation for the Generation of Energy', Int. Colloq. on Wind Energy, Brighton, British Wind Energy Assoc., 1981.

CHAPTER 10

SOME PRACTICAL HEATING
APPLICATIONS

INTRODUCTION

The previous chapters have shown that it is possible to use solar energy to
provide a proportion of the total heating demand in many parts of the world.
In high latitudes, however, it is important to appreciate that for many days
in the winter months the intensity of the solar radiation is too low to
provide any useful heat. There are two main types of solar installation
which can be tackled by a competent handyman who has had some experience
with basic carpentry and, preferably, some knowledge of standard water pipe
fittings. The simplest type of solar installation to construct is for low
temperature rise applications, such as in swimming pool heating. Here the
requirement is for large areas of simple, unglazed, uninsulated collector.
The second type of installation is more ambitious, as it involves the
domestic water heating system. The collector panels, which would typically
have an area of between 4 m^2 and 6 m^2 must be glazed and insulated and there
are several other factors to be considered, such as the relative positioning
of the various components and the length of the pipe runs. Solar
distillation and the construction of low-cost concentrators can also be
tackled with simple tools.

SWIMMING POOL AND OTHER LOW TEMPERATURE APPLICATIONS

An 'Enclosed' Collector

Although in high latitude countries such as the UK the simple, low temperature
rise enclosed collector would normally be used for the summer months only, it
operates at a high efficiency during this period and is very cost-effective.
The capital costs of these systems, excluding labour charges, would be
recovered in less than three years when compared with the anticipated
savings from most conventional sources. One enclosed collector design which
has been tested for over ten years is shown in Fig. 10.1. There are no glass
or transparent covers needed, as the temperature rise across the heater is
kept as low as possible. Provided the panels are located in a fairly
sheltered position they will perform at least as well as a glazed panel,
because there is always a radiation loss of approximately 10% in passing
through any transparent cover. There is no insulation provided at the sides

or the back, again because the temperature in the panel is normally close to ambient temperature and heat losses will be negligible. They are called 'enclosed' as the heated water flows underneath the heat absorbing material and does not evaporate.

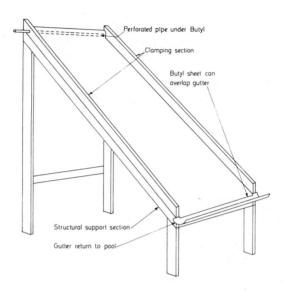

Fig. 10.1. Basic low-cost collector.

Structure

The main structural member of the panel is the backing sheet which can be based on any appropriate flat surface such as plywood sheet, preferably waterproofed. The standard 8 × 4 ft plywood sheet has proved easy to work with and provides an area of just over 3 m^2. The most important feature common to all these low temperature collectors is a thin matt black heat absorbing surface which can absorb nearly all the incident solar radiation. Butyl has proved to be a very satisfactory material for this application and the Butyl used on the original low temperature panels which I developed in 1968 (1, 2) showed no signs of degradation in 1980. This black surface sheet is placed on top of a second 'water flow spreading' surface so that the water which is to be heated can flow, under gravity, in a thin layer between the two sheets.

There are various ways of producing a thin uniformly spreading film of water on a sloping surface, but one method which has proved successful is to make the second surface out of a commercial polythene packing material known as Airwrap. This consists of a uniform matrix of equally spaced cylindrical air bubbles enclosed in polythene. The major disadvantage of this material is that its resistance to UV degradation is poor and it has a very limited life if exposed directly to solar radiation. However, when protected by the Butyl it has also survived for over ten years. The water inlet at the top of the collector consists of a small bore perforated pipe. The minimum pipe diameter should be 15.0 mm and the holes should be at least 2 mm diameter

and spaced about 10-15 mm apart. None of these dimensions are very
critical and it is easy to test the pipe before final assembly to check that
a uniform flow is obtained. A straightforward series connection of several
panels may not be completely satisfactory, as the pressures and rate of flow
in the system could mean that progressively less water reached successive
panels. This can be overcome either by a branching system, taking the
incoming water to each end of each panel or by increasing the flow area in
the panels with insufficient water by drilling more holes and/or enlarging
the holes. The heated water is returned to the pool by gravity, so that the
bottom of the panels must be higher than the pool surface. Plastic rainwater
gutters make excellent return channels and the evaporation loss is negligible.
Again, it is easy to test that the slope from the bottom of the panels to
the pool is adequate for the necessary flow.

List of Materials

(1) Flat panel for backing - $\frac{3}{8}$ in. weatherbonded ply is satisfactory
     (length L, width W).
(2) Butyl sheet.
(3) Airwrap packing sheet.
(4) Inlet pipe - plastic, 15 mm diameter is adequate. Length as needed for
     connecting to adjacent panels.
(5) Plastic rainwater gutter for return flow to pool. Length as needed for
     connecting to adjacent panels and return flow to pool.
(6) Structural supporting sections, two of length L, three or four of width
     W. Cross-section not critical, but adequate for rigidity.
(7) Top clamping sections of length L, primarily for sealing the edges.
(8) Framework to support panel, as required.
(9) A stop valve and appropriate flexible piping to connect with panel inlet
     pipes.

Lengths L and W could be a nominal 8 × 4 ft.

Brief Construction Details

The airwrap polythene is stretched over the plywood with the bubbles facing
upwards and held in place at the turned-over ends by a few drawing pins.
The butyl is laid over it and clamped by the section (7) above to the
structural support, as shown in Fig. 10.2. The main structural members are
fitted, then the perforated pipe. It is useful to test the pipe at this
point before fixing the top end of the butyl sheet to overlap it, to ensure
that an even flow of water is obtained. The return flow gutters should be
fitted last when the panels are sited, as it is necessary to have them sloping
gently towards the pool.

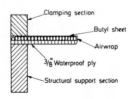

Fig. 10.2. Detail of clamping on sides.

Flow Rate

The temperature rise must be kept as low as possible to reduce heat losses.
One litre of water heated from 15 to 55°C has received only one tenth the
amount of heat supplied to 400 litres of water heated from 15 to 16°C,
although the former would be immediately detected and pronounced hot! Flow
rates should therefore be at least 150 litres/hour/m^2 of collector or about
3 UK gallons/hour/ft^2 of collector. It is important to ensure that the
collector panel is not distorted and that the entry pipe is horizontal.
Failure to check these points could lead to the water 'streaming' to one side
or the other, greatly reducing the overall effectiveness of the system as
heat from the solar radiation can only be transferred efficiently to the
water when it is directly in contact with the butyl sheet. A separate pump
is often not needed as the collector system can be connected by a T-pipe to
the return pipe from an existing filtration system.

Position and Direction

The angle of tilt and the direction which the panel faces are not very
critical. South facing panels are considered ideal for the northern
hemisphere, but a few degrees either side of south will make very little
difference. For the summer heating season only, a fairly well sloped panel
is best, perhaps 40° to the horizontal or less. It is possible to provide
complete computer simulations for the various periods of the year to predict
the optimum angle of tilt, but local conditions often impose greater
limitations, e.g. the presence of large screening trees or buildings. The
use of the roof of an existing building to mount the collectors will often
be very convenient and can reduce the visual impact of a large expanse of
black surface. Common sense should indicate any unsuitable, partially
shaded location, but it is advisable to check for over-shading on any site
over a full day in the early part of the heating season. The author has
seen at least two installations which were shaded from direct sunlight for
the greater part of the afternoon.

Size, Performance and Economics

The ratio of the collector area to the pool surface area is a convenient
ratio to start with. This ratio has been used (3, 4) to estimate the likely
daily temperature rise in a pool under various radiation conditions. To
obtain a temperature rise of about 5°C in one day under good summer conditions
in a temperate climate a collector/pool surface area approaching 1.5:1 has
been suggested, but it is important to appreciate that a steady increase in
the pool temperature over a period of several weeks early in the season can
be achieved with a ratio as low as 0.25:1. This is because the ground
surrounding the pool is also heated by the relatively higher temperature
pool water and this helps to maintain the pool at comfortable swimming
temperatures during periods of several consecutive cloudy days. Even with a
solar collector area of one tenth the pool surface area, sufficient energy
could be collected in one good day to give an additional temperature rise of
about 0.5°C. Tests carried out during the 1975 swimming season at a school
in Sussex, where panels based on the author's design principles were
installed showed that very substantial savings had been achieved compared
with the previous season. In 1974, with electric pool heating only, the
swimming season lasted from the end of May until the first week of September
and 48 885 units of electricity (kWh) were used (5). In 1975 with the solar
panels fitted to supplement the electric heating, the swimming season lasted

from mid-May until October, but the electricity consumption was reduced to
14 232 units. Average pool temperatures between 23 and 29°C were obtained,
with an estimated contribution of 500 kWh/m^2 from the collectors.

The cost of materials for the butyl, Airwrap and wooden-framed solar water
heater is less than £25/m^2 (at 1981 values) and this gives a payback period
of about two years on the basis of savings in heating costs compared with
conventional methods.

Other Designs - The 'Open' System

With a convenient south-facing (in the northern hemisphere) corrugated roof,
such as galvanized iron, a perforated pipe can be laid along the ridge and
the poolwater can be pumped up to the pipe and allowed to flow down the
corrugations. In these systems the flowing water is open to the atmosphere
and some heat losses due to evaporation are inevitable, making their overall
efficiency perhaps only two-thirds that of the enclosed type. The holes in
the perforated pipe should be placed opposite each groove in the corrugated
sheet with a minimum diameter of about 5 mm, as the distance between holes
will be at least 75 mm. Similar flow rates to the enclosed system should be
used to keep the temperature rise low. The efficiency of this system can be
improved by stretching a clear plastic material over the corrugations,
turning it into a type of Thomason system. Alternatively, a standard clear
outdoor corrugated plastic can be used.

Another 'open' system which has been tried successfully in Sussex (6)
consists of a large flat black stepped concrete area. The pool water is
pumped to the top step and cascades gently over the black concrete to the
pool. This is a very simple system, easy to construct and relatively large
collector/pool surface area ratios can be achieved. The only difficult
feature in the construction is to obtain a uniform thin film of water over
the entire surface area. The use of a long flexible perforated pipe at the
top step helps to achieve this.

Controls

Sensitive differential-temperature on-off control systems are now relatively
inexpensive. However, for these low temperature rise applications it has
been perfectly adequate to control the system manually, allowing the water
to flow through the panels from about 08.00 to 18.00 hours every day except
when it is very cloudy and overcast. If a differential-temperature
controller is used, it should have some type of time-delay in the circuit,
so that continuous cycling does not occur in intermittent cloud conditions.

Pool Cover

Before constructing a swimming pool solar heater remember that it is much
easier and far more cost-effective to provide a pool cover to reduce the
greatest source of heat loss from the pool - evaporation. The simplest
method is to use some type of floating cover. Light-gauge black polythene
sheet will make a difference. It can be clamped round the edges of the pool
and needs only small holes at approximately 0.3 m intervals to allow surface
rain water to pass through. Commercial floating pool covers are often made
of two layers of blue PVC, separated by strips of polyurethane foam.
Overnight temperature drops are usually in the order of 1°C compared with

over $2°C$ for the uncovered pool. This difference of $1°C$ sounds very small, but represents over 100 kWh in a small, 20 000 gallon pool.

Solar Distillation

In the simplest types of solar still, solar energy passes through a sloping transparent sheet, often glass, and heats saline water in an enclosed container. This causes the evaporation of some water and increases the humidity near the water surface. The warmer humid air circulates to the cooler transparent sheet where some of the water vapour condenses on the inner surface and slides down to be collected in a trough or container. Many different types of still have been reported in the literature (see also Chapter 6) and a useful practical description of some stills tested at the University of California has been reported by Howe and Tleimat (7), who include several diagrams. An even simpler approach for producing drinking water in arid regions has been used in Iran (8). A circular hole, about one metre in diameter, was dug in the ground and a beaker placed centrally in the hole. A transparent plastic sheet was then placed over the hole and covered round its edges by a layer of soil to prevent vapour leakage. The sheet was then gently pushed down in the centre and loaded with a small stone. No saline water from external sources needs to be supplied as capillary action and the temperature gradient in the soil are sufficient to cause the water vapour to form. This condenses on the plastic sheet and runs down to be collected in the beaker. Approximately one litre of fresh drinking water per square metre of collecting surface was obtained in the desert, some 500 km from the sea, during the first 24 hours of opertion, but the output fell off during the next ten days to about 150 cm^3.

Low-Cost Concentrators

A simple method of fabricating paraboloidal dish-shaped reflectors, which could be particularly useful for rural regions without access to sophisticated workshop facilities, has been described by Srinivasan *et al.* (9) who used aluminized mylar pasted on a suitable backing, such as cardboard, papier-mache, tinned or galvanized iron or thin aluminium sheets. A shape somewhat similar to an eight petalled flower is cut out, but the sides are based on the geometry of the parabola. When assembled, a dish-shaped reflector is obtained.

The Georgia Tech Spiral Concentrator (10) is based on Fresnel principles and is formed by slightly coiling a spiral cut from a flat sheet of material and attaching it to a simple planar frame at selected points. By 1981, nine Spiral Concentrators, ranging in concentration ratio from 50 to 500, had been fabricated with diameters up to 1.1 m. Materials used included electropolished anodized aluminium sheet and aluminium foil covered hardboard.

DOMESTIC SOLAR WATER HEATERS

There are probably hundreds of differert types of solar water heater design,
all of which make some contribution to the hot water requirements of their
particular installation. Most designs, however, have certain essential
features in common and these are as follows:

(1) The collector plate.
(2) Insulating material at the back and sides of the plate.
(3) One or two sheets of glass or translucent plastic in front of the
     collector plate.
(4) A casing to contain items (1), (2) and (3).
(5) A hot water storage system, which may be a separate storage tank.

The major British work in this field was carried out by the late Professor
Harold Heywood, between 1947 and 1955, and his design principles have
subsequently formed the basis for many types of solar heating system
(11, 12).

Towards the end of the 1970s a very practical text by McCartney and Ford (13)
and the Building Research Establishment's guide to the design, installation
and economic aspects of solar heating systmes for the UK (14) were published.
These books are complementary and give very detailed advice.

Casing, Covers and Insulation

The purpose of the casing and glass or plastic cover(s) is to act as a
weatherproof container for the collector plate. Any conventional flat box-
like shape can be modified to hold the collector plate and support the
cover(s). Glass reinforced plastic (GRP) is a fairly popular choice for the
casing, but wood or sheet metal can also be used. Examples of typical cross-
sections illustrating each type are shown in Fig. 10.3.

A single sheet of glass will transmit about 90% of the incident solar
radiation but will trap nearly all the heat radiated by the collector surface,
as glass is opaque to longwave radiation. The use of two covers further
reduces the amount of radiation which reaches the collector plate, but if the
collector plate is more than about 35° above the temperature of the surroundings, a second cover improves the collector efficiency as it reduces the heat
losses from the outer cover to the surroundings. It also helped to protect
the plate from overnight freezing temperatures in winter. The increased cost
and difficulty of fitting a second cover, as well as the comparatively small
performance advantage obtained in UK conditions, make a single cover the
recommendation for the simple practical collector system.

As an alternative to glass, the use of translucent plastic can be considered,
provided that it has been specially treated to stand up to the weather.

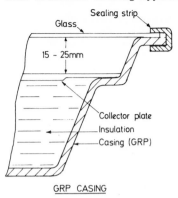

GRP CASING

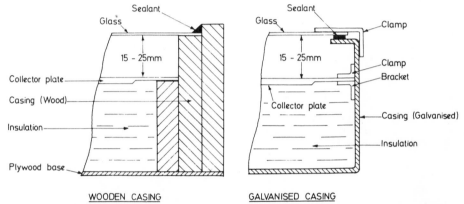

WOODEN CASING   GALVANISED CASING

Fig. 10.3.

Early weather resistant materials from DuPont, Mylar and Tedlar PVF type
400 BG20 TR, have been replaced by Tedlar type 400 SE PVF film. This is a
100 μm film which can withstand temperatures up to 180°C and can be heat
sealed, shrink wrapped or bonded by adhesives. Glass-reinforced translucent
plastic (GRP) sheets have been used successfully both in the UK (15) and the
USA (16). GRP is easier to handle than glass, especially if working on an
exposed roof as considerable care has to be taken with glass to avoid
breakages. The distance between the parallel surfaces of two covers or a
single (or inner) cover and the collector plate should be between 15 and
25 mm. The exact spacing is not very critical. For cheapness, with
comparatively little loss in overall performance, 4 mm horticultural grade
glass can be used for the cover. One reservation about the use of plastic
materials is that even those which have been specially developed to withstand
weathering will have a limited lifetime and the manufacturers should be
consulted directly about the likely life. In designing the detail of fitting
and sealing the cover to the casing, avoid leaving a water trap at the edges.
Some designs have ignored this and consequently have a small pool of dirty
water almost permanently on the lower edge of the collector cover surface.
Various commercial insulating materials can be used at the rear and sides of
the collector plate, providing they can stand up to maximum temperatures of
over 100°C - quite possible with the system not operating and exposed on a

hot sunny day. Fibreglass or mineral (rock) wool is very satisfactory but
polystyrene should not be used as an insulating material, as it melts if
placed in close contact with a hot collector plate. A minimum thickness of
50 mm of insulation is suggested at the rear of the collector plate, with a
minimum of 25 mm at the sides although this is not so critical and could be
omitted.

COLLECTOR PLATE DESIGNS

General

Selective surface. The provision of a selective surface is beyond the scope
of the great majority of practical home workshops. Copper is probably the
easiest material to coat (17). Even with the commercially available
coatings, there are conflicting views about how long they continue to be
effective. It can be seen from the performance curves in Chapter 3 that
selective surfaces have an advantage when the temperature of the collector
is relatively high with good radiation conditions, or at moderate temperatures
with average radiation. Adhesive backed foils are a useful alternative to
coatings.

Frost protection. The problem of freezing in the winter can be dealt with
in several ways. The simplest is to forget about solar heating for the
entire winter period and drain the collectors. The amount of heat not
collected between mid-October and mid-March would represent perhaps 20% of
the total annual potential energy gain. If an anti-freeze solution is used,
the system must be completely self-contained and should have the approval of
the Local Water Authority. Such systems use an indirect hot water tank
fitted with a heat transfer coil connected directly to the solar collectors
and are discussed later.

Corrosion. The problems of corrosion were also dealt with in Chapter 3.
One cause of trouble is likely to be a combination of mixed metals in the
system, such as copper and aluminium, either directly in contact under moist
conditions, or if an aluminium collector plate with hollow passages is
subjected to ordinary mains water containing certain dissolved chemicals.
Although there may be no direct contact between dissimilar metals in a solar
water heating system, corrosion problems may occur whenever copper and steel
or galvanized components are used, particularly when the water is cuprosolvent. In the early stages of use, copper from the collector plate or
connecting pipes could dissolve in the water and be redeposited in the
galvanized storage tank. Similarly a galvanized steel collector plate could
corrode if it were connected to a copper tank (18).

The presence of dissolved oxygen in the system is another equally important
factor in corrosion, but can be completely overcome by the use of an all-
copper system. Copper, which is very widely used in general plumbing
practice, does not corrode in oxygenated water or in a suitably treated
antifreeze solution.

Pipe runs. For normal installations operating on a thermosyphon system, a
pipe diameter of 28 mm is recommended between the collector and the storage
tank. Normal good plumbing practice should be followed in all pipe runs,
keeping right angle bends to a minimum, especially in thermosyphon systems.
The main problem likely to arise is the formation of air locks in the
system. When ordinary mains water is heated, dissolved air comes out of
solution and the slow accumulation of bubbles at any point establishes an
air lock which either completely stops the flow or reduces the circulation

rate. It is essential that there should be a continuous rise in both the
flow and return pipes from the collector to the storage tank. It is not
essential but can be advantageous for the collector to be slightly inclined
so that the horizontal header sections rise gently towards the collector
outlet. Adequate provision should be made for vent pipes in the system.
Pipe lengths between the various component parts in the system should be as
short as possible. All heated pipes should be insulated.

Specific

Corrugated galvanized collector. Heywood's first practical domestic
collector, installed in his home near London in 1948, consisted of two sheets
of corrugated galvanized steel placed in 'mirror image' position so that
eight water channels were formed along the length of the collector surface.
The edges were rivetted and soldered, and had square section headers fitted
at the top and bottom edges. The front surface of the collector, which had
an area of just under 1 m^2, was painted matt black and enclosed in a wooden
frame with two sheets of glass over the front and insulating material at the
back. He commented that while the collector worked well in a conventional
thermosyphon system for a number of years "it did not have a long life" (12).

Nevertheless, a modified version of the original Heywood collector has been
developed very successfully by the Brace Research Institute (19). The Brace
Collector was designed to incorporate low-cost materials generally available,
even in relatively remote parts of the world, and is based on two galvanized
steel 22 gauge sheets, one being corrugated and forming the heat absorbing
surface and the other being flat. The two sheets are rivetted and soldered
together, the corrugated surface is painted black and the collector is
placed in a simple galvanized steel box lined with an insulating material –
coconut fibre is suggested. A single sheet of 3 mm window glass is fitted
using a silicone sealant and allowing a 3 mm clearance all round for glass
expansion. Hot water storage is in a converted 45 gallon oil drum. An
initial life of about five years with negligible maintenance is claimed and
more than seven years satisfactory operation was reported for some units in
Barbados.

Pipes bonded to a metal sheet. The Commonwealth Scientific and Industrial
Research Organization (CSIRO), Australia, published a guide to the principles
of the design, construction and installation of solar water heaters in 1964
(17) which was summarized in the JIHVE in 1967 (20). The absorber plate
described in their guide consisted of pipes thermally bonded to a metal sheet.
Copper was their preferred material and a framework of 28 and 15 mm copper
pipes were soldered to a 26 gauge copper sheet (approximately 0.45 mm), the
vertically rising 15 mm pipes being brazed to the 28 mm header pipes.
Galvanized steel or asbestos cement was recommended for the collector
casing at that time. Slightly thicker copper sheet, of 20-24 gauge
(approximately 0.56-0.91 mm) is currently recommended in the UK. A plan view
of a typical pipe matrix is shown in Fig. 10.4. This could be connected or
bonded to a flat or corrugated sheet. The recommended distance between the
centre line of adjacent pipes is about 150 mm.

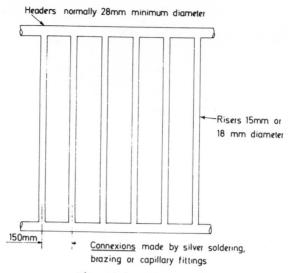

Fig. 10.4.   Pipe matrix.

Although copper is the preferred material, galvanized or aluminium pipes or sheet could be used. Any departure from the excellent thermal bond which can be achieved by the copper-copper system will be less efficient, the worst bond being achieved by simple wire tying at large intervals. The Henry Mathew collector (21) however, shown in Fig. 10.5, achieves good results with a theoretically poor collector, as the distance between wire ties is about 750 mm. In this all steel system, the galvanized pipe in the collector is horizontal, but adjacent pipes are much closer together than the recommended 150 mm.

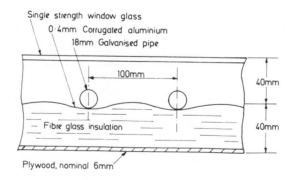

Fig. 10.5.   Section of Henry Mathew collector.

The use of clips at shorter intervals will improve the bond, and forming a semi-circular depression in a flat sheet to fit the pipe will also add to the performance, particularly if some form of adhesive or filler is then added, if soldering is not appropriate.

This is one of the approaches described by MacGregor (22) who also suggested improved methods for connecting groups of collector panels.

As an alternative to the matrix system, a continuous serpentine loop can be used, as shown in Fig. 10.6. This idea is used in some commercial panels, but its use is limited to forced circulation systems.

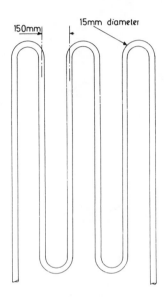

Fig. 10.6. Serpentine system.

Commercial steel panel radiators. The standard commercial steel panel radiator can be easily modified to make a solar collector plate. If possible, the panel should be obtained without its final coat of glossy white paint, as the absorbing surface should be painted with a standard matt black paint. The panel radiator which I first tested in 1968 still had its original black painted surface intact after eight year's exposure under a single Mylar transparent cover sheet (23). The rear side of the panel can be left white. There are normally four connections at the top and bottom of the panel at each end. The cold return water inlet should be arranged to enter a bottom connection and the solar heated water should leave from the diagonally opposite connection, i.e. bottom left to top right or bottom right to top left. Do not connect the inlet pipe by a branching connection to each end of the panel or try to take the heated water from both top outlets, as this can reduce the overall efficiency. In the thermosyphon system, for example, an internal circular flow pattern could be established with a double inlet and outlet arrangement. The painted radiator, with its connections ready, should be placed in the casing with its ribs following the normal pattern, i.e. running upwards to the horizontal header as shown in Fig. 10.7. The panel will work if turned at right angles, but with an overall loss of efficiency.

Fig. 10.7.

A simple trough collector/storage tank. An effective collector can be made
out of a watertight casing with a sloping bottom as shown in Fig. 10.8.
This is a combined collector and solar storage tank and is particularly
suitable where there is only an existing cold water supply. It will not be
suitable for poor radiation conditions or freezing ambient temperatures. The
cold inlet water displaces the heated water at the shallower end of the
casing when the inlet control valve is opened. The use of the sloping bottom
ensures that after a brief period of good radiation conditions there is a
layer of heated water available. A few small ventilation holes should be
drilled in the casing below the glass cover to minimize the effect of
moisture condensation. GRP painted matt black or lined with butyl is
suggested for the casing material. A simpler version consisting of a square-
sectioned uniform depth galvanized casing was described and tested by the
National Building Research Institute, Pretoria in 1967 (24).

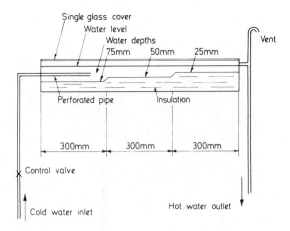

Fig. 10.8.   Flat trough collector/storage.

INTEGRATING SOLAR WATER HEATING INTO THE DOMESTIC HOT WATER SYSTEM

The Thermosyphon System

The basic components in a standard domestic system are shown in Fig. 10.9.
The simplest method of using solar water heaters is in the direct natural
thermosyphon system, shown in Fig. 10.10, using a separate solar hot water
storage tank. As the water is heated in the collector it rises to the top
of the collector and then passes to the upper section of the storage tank.
At the same time the cooler water at the bottom of the storage tank returns
to the bottom of the collectors. As the flow is caused by the difference
in density between the hot and cold water there must be a difference in
level, H, between the bottom of the storage tank and the top of the
collector. With a minimum of 600 mm it is unlikely that any reverse flow at
night could take place, but a non-return valve could be fitted in the cold
water flow pipe. Chinnery (25) reported that any reduction of H below 600 mm
led to a corresponding reduction in the overall efficiency of the system as
indicated in Table 10.1.

Table 10.1

	Mean efficiency
Collector outlet at 600 mm below bottom of tank	54.6%
Collector outlet level with bottom of tank	46.4%
Collector outlet about two-thirds tank height above bottom	43.8%

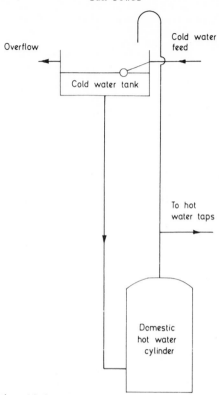

Fig. 10.9. A standard domestic system.

The hot water pipe from the collectors should enter the upper section of the
storage tank at a level lying between two-thirds and three-quarters of the
total capacity. Again, Chinnery shows that loss of efficiency occurs when
this entry point is lowered. The use of a separate storage tank has an
advantage in countries such as the UK because for the greater part of the
year the temperatures that can be achieved from the solar collectors are too
low for direct use in the domestic hot water system. It is useful to arrange
for the possibility of by-passing the hot water cylinder during good
radiation conditions in the summer so that conventional heating can be turned
off. This is particularly useful if no hot water is drawn off during the
day, as otherwise, in the evening, there could be a mixing of very hot solar
heated water with the unheated water in the hot water cylinder. With the
separate solar hot water storage tank, even a small temperature rise above
the cold water tank temperature can save energy, as this heated water enters
the hot water cylinder instead of the direct cold feed water. The thermosyphon system can also be used indirectly, a shown in Fig. 10.11. A heat
exchanger is fitted inside the solar hot water storage tank and there is a
closed circuit from the collectors through the heat exchanger. This
circuit contains a sealed expansion tank, or provision for a separate cold
water 'topping-up' tank and an overflow pipe. The sealed expansion tank
prevents fresh oxygen from getting into the system and inhibits corrosion.
The indirect circuit can be filled with an antifreeze solution, but this
must be completely self-contained so that the antifreeze does not leak into
the domestic system. A heat exchanger can be a simple coil of copper pipe
and some commercial groups recommend a standard small domestic copper

cylinder as the storage tank/heat exchanger unit. These are not designed
for operating at the lower temperature differences and flow rates likely to
be encountered in a solar installation and it is far better to use about one
metre of finned 28 mm diameter copper pipe for 200 litres of storage tank.
With indirect circuits care should be taken to connect the hot water pipe
from the collectors to the top of the heat exchanger. The system could work
with the hot water pipe connected to the lower end of the heat exchanger, but
it would be considerably less effective, as the water flow rates would be
much lower.

Forced Circulation or Pumped Systems

As can be seen from Fig. 10.12, a forced circulation system is more
complicated and the circulating pump, which can be a normal small central
heating pump, must be controlled by a differential-temperature controller.
These are supplied by various solar collector manufacturers or can be built
from an electronic circuit design. An analysis of the operation of a
differential-temperature controller and the type of problem encountered has
been given by O'Connel (26), who warns that on days with comparatively low
radiation intensity and intermittent cloud cover, a system could cycle
continuously, loosing more energy than is gained. The setting of the
temperature difference for switching on and off the pump is important, as is
the position of the temperature sensors, which should not be placed at a
high level on the storage tank.

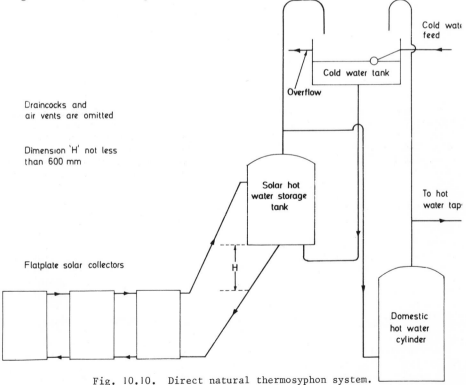

Fig. 10.10.   Direct natural thermosyphon system.

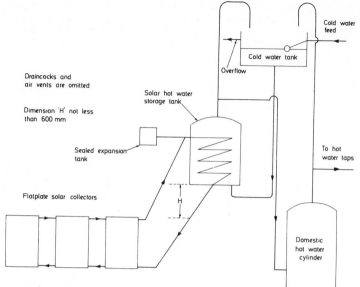

Fig. 10.11.   Indirect natural thermosyphon system.

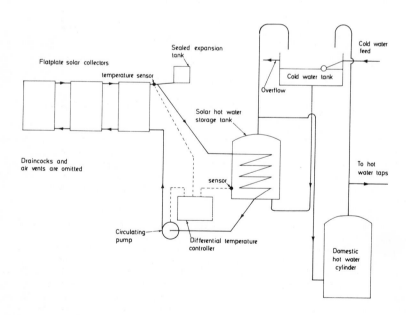

Fig. 10.12.   Indirect forced circulation system.

An alternative method of protecting the system against frost damage is
shown in Fig. 10.13, where the collectors can be completely drained by a
solenoid-operated drain valve. This system is more elaborate than the
others and care must be taken in deciding the height and position of the
various components, so that the cold water tank does not empty when the
drain valve is opened.

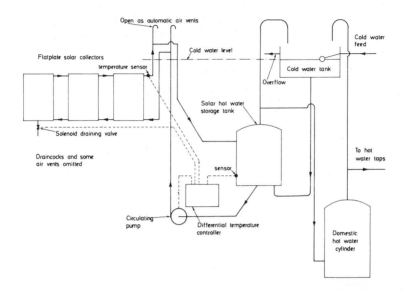

Fig. 10.13.   Direct forced circulation system with frost
protection.

A system which eliminates the separate solar hot water storage tank is shown
in Fig. 10.14. The combined domestic/solar cylinder has two heat exchanger
coils in it, the upper one being connected directly to a conventional boiler
system. The advantages are that less space is taken up and fewer pipes are
used, but under ideal conditions the efficiency will be less than with a
separate storage tank.

The entry of the cold water supply to the hot water cylinder can create
problems. If the cold water enters in a vertical direction it would mix
with the heated water near the top of the cylinder, cooling it very rapidly.
One way of overcoming this is to position the entry pipe so that it
discharges horizontally or slightly downwards into the tank. The cold water
supply should never be connected to the return pipe of a direct thermosyphon
system. In principle this could be satisfactory during the day, as the cold
water would absorb heat as it passed through the collector plate, but at
night the cold water would also enter the hot water cylinder near the top
and mix directly with heated water.

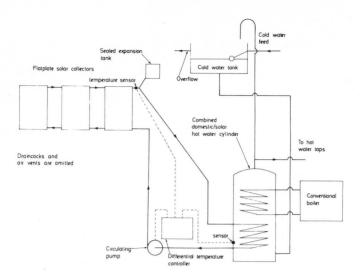

Fig. 10.14.   Single cylinder indirect forced circulation
            system.

In the UK a further modification suggested by Pallis (27) and Makkar (28)
is the addition of a third tap in the bath or sink which is connected
directly to the solar pre-heating storage tank. Preliminary results from
field trials in south London suggest an increase in useful solar energy by
between 20% and 30%, a figure which is supported by computer analysis.

                SIZE, PERFORMANCE, ECONOMICS AND STORAGE CAPACITY

The British Standards Institution Code of Practice (BS 5918:1980) was the
first UK standard to deal with the direct use of solar energy and was
intended to help the developing solar industry to provide well constructed
and durable equipment for pre-heating domestic hot water. It also provided
guidelines for estimating the likely performance from different collector
systems when operating under UK conditions. One of the interesting features
in the Code was that system performance was unlikely to vary by more than
10% from an annual mean value provided the following ranges were adopted:

(1)   Geographical location in England and Wales.
(2)   Collector orientation to be between south-east and south-west.
(3)   Collector slope to the horizontal to be between 5° and 60°.
(4)   Collector thermal capacity to be less than 9 litres of water (and water
      equivalent) per square metre.

This is shown in Fig. 10.15 which refers to a single glazed matt black
collector. System performance is very sensitive both to the actual usage of
hot water and to the installed collector area.

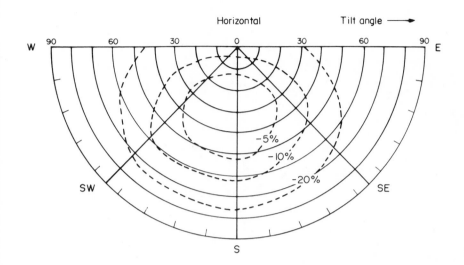

Fig. 10.15.   The variation of solar energy contribution
with orientation and tilt for a typical
solar water heating system in the UK.

Heywood's work carried out during the 1950s and 1960s (29) showed that in
the UK at a latitude of 51°31' North, a south facing surface inclined at an
angle of 40° to the horizontal would receive a daily average of 9.2 MJ/m²
or about 2.56 kWh/m².

He was also the first to point out that differences of a few degrees in the
angle of inclination or in direction from south make very little difference
to the total annual performance of a solar collector system.

The British Solar Code bases its performance figures on a reference system
in which a direct system with a preheat vessel was linked to a collector
area of 4m², facing south and inclined at 45° to the horizontal.   The preheat
vessel had a capacity of 200 litres and the domestic hot water draw-off was
175 litres per day at a temperature of 60°C.   The cold water feed temperature
varied throughout the year between 6°C and 12°C.   Five classes of collector
are defined, with single-glazed being in class IV or V, double-glazed usually
in class IV, and single glazed with selective surfaces in class III.
Evacuated tubular collectors would fall in class II or I.   System performance
is given in Table 10.2.

Table 10.2

Collector class	Solar energy supplied per year by reference system (collector area 4 m^2) with daily drawoff of 175 litres at 60°C	
	MJ	kWh
I	more than 7000	more than 1944
II	6000 to 7000	1677 to 1944
III	5000 to 6000	1389 to 1667
IV	4000 to 5000	1111 to 1389
V	less than 4000	less than 1111

Note: performance obtained with the amount of solar radiation recorded at Kew for the year November 1963 to October 1964.

This agrees very closely with my earlier assessment in which it was stated that laboratory tests can give values of collector efficiencies well over 60% for moderate temperature differences, but when due allowance is made for the longer pipe runs in an actual installation, the intermittent nature of the sunshine during the day and the way in which water is used, an overall figure of between 30% and 40% is realistic for the UK (30). This means that an annual total of between 280 and 376 kWh can be obtained per square metre of collector. The figure of 280 kWh was confirmed in a series of tests carried out for the year September 1973 − August 1974 (31), although − this was regarded as a low figure for two reasons — it was a below average year for solar radiation and the storage tank was not well insulated. The Building Research Establishment have quite independently suggested (32, 33) figures of 324 kWh/m^2 for a 6 m^2 installation and 350 kWh/m^2 for a 4 m^2 installation. There is an important feature to appreciate about these overall values for efficiencies between 30% and 40%. Any increase in collector area above 6 m^2 for an average household will not give a proportionate increase in the total amount of useful heat collected. If it did, then an area of about 12 m^2 would provide hot water all though the year. This is impossible because of the very low levels of radiation received in the winter period. To approach the average daily demand of about 10 kWh towards the mid winter period, a collector area of about 50 m^2 would be necessary, but this would not work, as there is another practical barrier − all flat plate collectors have some lower limit of radiation level below which they cannot operate. The best figures to take should be based on at least 3 m^2 for two adults, and 4 m^2 for the family of three or four, with system performance based on Table 10.2 annual savings can then be calculated by estimating the performance, taking account of a 'third tap' if appropriate, and using the cost of replaced fuel per unit saved. For example, if electricity at 5p per kWh is replaced by a single-glazed collector system with an area of 5 m^2 and a selective surface, this could save 1750 kWh per annum, or £87.50.

The ratio of approximately 50 litres of hot water storage capacity for every square metre of collector area (one UK gallon/ft^2), originally suggested by Heywood, has been accepted as the standard for the great majority of domestic solar heating systems.

LOCAL BUILDING AND PLANNING REGULATIONS

Solar heating systems must comply with local building and planning regulations. For example, if a collector is erected on a roof or fixed to a house it must be secure and not liable to be blown off in a high wind. It is also probable that some planning authorities would raise objections if solar heating panels made a substantial alteration to the visual appearance of a building. This would be particularly relevant in the case of older buildings of historic interest. Many people might object to the rather stark appearance of swimming pool heaters and it may be necessary to site these behind a hedge or similar screening in conditions which are somewhat less than ideal.

POINTS TO BE CHECKED

During 1979 the UK Solar Trade Association published a code of practice for the industry (34). Their check list, shown below, provides a valuable guide to the main points which need to be checked in any installation.

UK Solar Trade Association Check List

(1) All unions and glands are free from weeps.
(2) The glazing seals are weathertight and sound.
(3) All air has been expelled from the system.
(4) The levels in the systems are correct or sealed systems are at the correct pressure.
(5) Electrical controls and temperature sensors are operating correctly.
(6) The circulating pump, if fitted, is operating without undue noise.
(7) All insulation is firmly in place.
(8) All covers are properly in place.
(9) No condensation or damp spots are apparent, particularly around the pipes and fixings in the roof space.
(10) The roof fixings are firm and the roof covering satisfactory by visual inspection.
(11) The heat transfer fluid is the correct type.
(12) Safety devices are operating properly.

THE CLAIMS OF COMMERCIAL BROCHURES

While the solar industry was developing in the UK during the 1970s, some misleading statements were published. Typical examples of these statements were: "solar heat can provide nearly all your domestic hot water requirements free" and "the panels will heat all the domestic water needed by an average family in the summer and 80% of the required water in the winter". This is not necessarily untrue, but it could only happen if people were prepared to alter their way of life quite radically and face the prospect of storing dirty dishes, cups and saucers, dirty clothes etc. for weeks on end during the winter months while waiting for a few sunny days. Even in the summer months there are often several consecutive cloudy days with low solar radiation levels when very little heat can be transferred to the domestic water heating system. Some manufacturers do provide useful independent test results and these can be plotted on the characteristic performance chart for single and double covers, shown in Fig. 10.16. The chain-dotted line was calculated from the information given in the following extract from a UK brochure (Senior Platecoil Ltd):

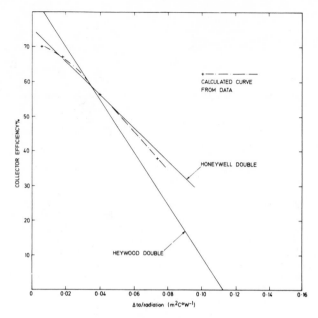

Fig. 10.16.

When NASA Lewis in Cleveland tested the Solar Collector the conditions were:

(1)   300 Btu/hr/sq ft input flux level.
(2)   7 mph wind.
(3)   2 glasses of $\frac{1}{8}$ inch green glass with 88% transmission level.
(4)   10 lb/hr/sq ft water flow rate.
(5)   80°F ambient (temperature).
(6)   Plain brushed on black paint.

The results, considered to be good, were as follows:

Inlet water temperature (°F)	Efficiency (%)	Heat pick up (Btu/hr/sq ft)
80	70	210
100	67	201
140	56	168
200	38	114

Can their claim that "the results were considered to be good" be substantiated? First, the mean collector plate temperature has to be calculated. The flow rate is given, so is the heat pick up, and the mean temperature rise across the collector plate, Δt_c, is heat pick up divided by flow rate. The mean temperature difference of the collector plate and the surroundings, Δt_a, is given by $(Ti - 80) + \Delta t_c/2$. The final stages in the calculation are to divide the mean temperature difference, Δt_a, by the total radiation landing on the collector and then convert the units into °C W^{-1} m^2. The new table is then as follows:

Table 10.3

Inlet water (Ti) $(^{\circ}F)$	Efficiency (%)	Δt_c $(^{\circ}F)$	Δt_a $(^{\circ}F)$	Δt_a 300	$(^{\circ}C\ W^{-1}\ m^2)$
80	70	21	10.5	0.035	0.0061
100	67	20.1	30.05	0.010	0.0176
140	56	16.8	68.4	0.228	0.0401
240	38	11.4	125.7	0.419	0.0738

The points joined by the chain-dotted line on Fig. 10.15 lie very close to the Honeywell double glazed collector performance and are better than Heywood's early work, so the manufacturer's claim that the results were considered to be good is quite justified.

By the early 1980s most collectors were being tested in this way. The claims of a few manufacturers and installers to have achieved annual savings in the UK approaching 1000 kWh per m² of collector area in a domestic hot water system have never been verified and, in some cases, have resulted in successful prosecutions by various County Trading Standards Departments.

OTHER ENERGY SAVING METHODS

Although the provision of adequate loft insulation and trying to eliminate draughts by sealing round the edges of windows and doors will not be as interesting or exciting as building a solar water or space heating system, these simple measures will be far more cost effective at present. An analysis carried out in the UK by the author in 1981 for a four-bedroomed semi-detached house gave the following figures for capital cost and estimated fuel savings over a five year period:

	Cost of installation	Estimated value of saved fuel in 5 years (with no inflation)
Basic roof insulation (100 mm)	£125	£300
Draught prevention	£25	£100

This should be compared with solar heating at 1981 prices

	Cost of installation	Estimated value* of saved fuel in 5 years (with no inflation)
Practical 5 m² system (excluding labour)	£400–£500	£350
Commercially available 5 m² system	£1000–£1400	£350

---

*Compared with normal electricity tariff.

REFERENCES

(1) Dewhurst, J. and McVeigh, J. C., A low-cost solar heater, *The Heating and Ventilating Engineer*, March 1968.

(2) McVeigh, J. C., Some experiments in heating swimming pools by solar energy, *JIHVE* 39, 53-55, June 1971.

(3) How to heat your swimming pool using solar energy, Brace Research Institute, McGill University, January 1965, revised February 1973.

(4) deWinter, F. and Lyman, W. S., Home built solar water heaters for swimming pools, ISES Congress 'The Sun in the Service of Mankind' Paris, 1973.

(5) Plumb, M., Solar tanning for swim pool heating bill, *Sussex Express and County Herald*, 28 May 1976.

(6) Carter, The Hon. Mrs B., Brencar Solar Exports Ltd, Rogate, Hants.

(7) Howe, E. D. and Tleimat, B. W., Twenty years of work on solar distillation at the University of California, Solar Energy, 16, 97-105, 1974.

(8) Ahmadzadeh, J., Solar Earth-water stills, *Solar Energy*, 20, 382-391, 1978.

(9) Srinivasan, M., Kulkarni, L. V. and Pasupathy, C. S., A simple technique of fabrication of paraboloidal concentrators, *Solar Energy* 22, 463-465, 1979.

(10) Steenblik, R. A., The development of a simple and innovative low cost concentrator, Paper C2:28, ISES Solar World Forum Abstracts, Pergamon Press, Oxford, 1981.

(11) Heywood, H., Solar Energy: Past, present and future applications, *Engineering* 176, 409, 1953.

(12) Heywood, H., An appraisal of the use of solar energy, *Society of Engineers* 57, 155, 1966.

(13) McCartney, K. with Ford, B., *Practical Solar Heating*, Prism Press, Dorchester, 1979.

(14) Wozniak, S. J., Solar heating systems for the UK: design, installation, and economic aspects. Building Research Establishment Report, HMSO 1979.

(15) Brachi, P., Sun on the roof, *New Scientist*, 19 September 1974.

(16) Scoville, A. E., An alternative cover material for solar collectors, ISES Congress, Los Angeles, Extended Abstracts, Paper 30/11, July 1975.

(17) Solar Water Heaters, CSIRO Division of Mechanical Engineering, Circular No. 2, 1964.

(18) McVeigh, J. C., Advances in Solar Energy, *Heating and Ventilating News*, September 1975.

(19) How to build a solar water heater, Brace Research Institute, McGill
      University, February 1965, revised February 1973.

(20) Solar Water Heaters, a summary of (12), *JIHVE* 309, January 1967.

(21) Reynolds, J. S. *et al.*, The Atypical Mathew Solar House at Coos Bay,
      Oregon, ISES Congress, Los Angeles, Extended Abstracts, Paper 43/12,
      July 1975, and in *Solar Energy* 19, 219, 1977.

(22) MacGregor, A. W. K., The Collector Plate, UK ISES Conference (C13) on
      Practical Aspects of Domestic Solar Water Heaters, 1977.

(23) McVeigh, J. C., Developments in solar energy utilisation in the United
      Kingdom, ISES Congress, Los Angeles, Extended Abstracts, Paper 10/4,
      July 1975, and in *Solar Energy* 18, 381-385, 1976.

(24) Richards, S. J. and Chinnery, D. N. W., A solar water heater for low-
      cost housing, National Building Research Institute, Bulletin 41, CSIR
      Research Report 237, Pretoria, South Africa 1967.

(25) Chinnery, D. N. W., Solar Water Heating in South Africa, National
      Building Research Institute, Bulletin 44, CSIR Research Report 248,
      Pretoria, South Africa, 1967.

(26) O'Connel, J. C., The problems associated with the use of differential
      temperature controllers, Solar Energy for Buildings Seminar, North
      East London Polytechnic, February 1976.

(27) Pallis, S., The Solar bath-tap, *Solar Energy* 25, 531-536, 1980.

(28) Makkar, L., Solar assisted domestic hot water: three tap systems,
      Solar World Forum, Vol. 1, 86-93, Pergamon Press, Oxford, 1982.

(29) Heywood, H., Operating experiences with solar water heating, *JIHVE* 39,
      63-69, June 1971.

(30) McVeigh, J. C., *Sun Power*, First Edition, 177, Pergamon Press, Oxford,
      1977.

(31) Harris, J., The British solar panel is born, *Building Services Engineer*
      42, 432, October 1974.

(32) Building Research Establishment, Energy Conservation: a study of
      energy conservation in buildings and possible means of saving energy
      in housing, BRE Current paper CP 56/75, 1975.

(33) Courtney, R. G., An appraisal of solar water heating in the UK, BRE
      Current paper CP 7/76, 1976.

(34) Solar Trade Association Ltd, Solar Water Heating Code of Practice,
      STA, 26 Store Street, London WC1E 7BT, 1979 and second edition
      July 1982.

APPENDIX 1
SOME USEFUL UNITS, DEFINITIONS
AND CONVERSIONS

Système International (SI) units have been adopted by many countries including the United Kingdom. Some papers on solar energy, particularly the earlier references, use other systems of units. The basic SI units are as follows:

metre	m	length
kilogramme	kg	mass
second	s	time
kelvin	K	thermodynamic temperature

PREFERRED ABBREVIATIONS IN SI UNITS

Tera	T	10^{12}	milli	m	10^{-3}
Giga	G	10^9	micro	μ	10^{-6}
Mega	M	10^6	nano	n	10^{-9}
Kilo	k	10^3	pico	p	10^{-12}

The following definitions and conversion factors may be used:

LENGTH

1 millimetre (mm)	0.0393701 inch (in.)
1 metre (m)	3.28084 ft
1 Ångström (Å)	10^{-10} m

AREA

1 square centimetre (cm^2)	0.155000 in^2
1 square metre (m^2)	10.7639 ft^2
1 hectare	10^4 m^2 or 2.471 acres

VOLUME

1 cubic centimetre (cm^3)	0.0610237 in^3
1 litre	10^3 cm^3 or 10^{-3} m^3
1 Imperial gallon (UK)	4.54596 litre
1 US gallon	3.78531 litre

MASS

1 kilogramme (kg)	2.20462 lb
1 tonne (10^3 kg)	0.984207 ton (UK)

The basic unit of force is the newton (N). One pascal (Pa) is exactly one newton per square metre (N/m^2).

$$1 \text{ lbf} = 4.448 \text{ N}$$

PRESSURE

1 bar	10^5 Pa
1 atmosphere	101.325 kPa
1 lbf/in^2	6.894 kPa
1 kg/cm^2	14.2233 lb/in^2

HEAT, ENERGY AND POWER

The British Thermal Unit (Btu) is the amount of heat required to raise 1 lb of water through 1 degree Fahrenheit.

The calorie is the amount of heat required to raise 1 g of water through 1 degree celsius (Centigrade).

The Langley is a unit of energy frequently used in radiation work and is equivalent to 1 calorie/cm^2.

Heat is a form of energy and the joule (J) is commonly used as a mechanical unit of heat. The fundamental unit of power is the watt (W).

$$0^\circ C = 273.16 \text{ K} = 32^\circ F$$

1 Btu = 1.05506×10^3 joule (J) = 778.169 ft lb

1 calorie (cal) = 4.1868 J

1 kilocalorie = 3.96830 Btu

1 Watt = 1 Joule per second = 0.00134 horsepower

1 kilowatt hour (kWh) = 3.600×10^6 J = 3.600 MJ = 3.41213×10^3 Btu

1 Btu/h = 0.293071 W

1 kilocalorie/m^2 = 0.368668 Btu/ft^2 = 1.163 W h/m^2

1 W/m^2 = 3.6 kJ/m^2/h = 0.316998 Btu/ft^2/h

1 Btu/h ft^2 $^\circ F$ = 5.67826 W/m^2 $^\circ C$

Appendix 1

SOME APPROXIMATE THERMAL ENERGY EQUIVALENTS OF FUELS

Oil   1 tonne = 4.48×10^{10} J
Black Coal 1 tonne = 2.9×10^{10} J
Natural Gas 1 ft^3 at STP = 1.05 MJ, or 1 m^3 = 37.1 MJ

APPENDIX 2

UK OBSERVING NETWORK AND THE STORAGE OF DATA

The primary solar radiation network in the UK is the responsibility of the Meteorological Office. Monthly mean values of global radiation, diffuse radiation and illumination on a horizontal plane appear fairly soon after observation in the Monthly Weather Summary. These data are provisional data, and are subject to slight revision after appropriate calibration and consistency checks. The Radiation Section at the Meteorological Office processes the observed data, and when they are satisfied, hourly data are transferred onto magnetic data tapes containing radiation, illumination and sunshine data only, based on local apparent time. Daily totals are summarized on published data sheets which appear about two to three years after actual observation.

Enquiries for more detailed information about the UK radiation network and the associated data banks should be addressed to the Meteorological Office, Eastern Road, Bracknell, Berkshire RG12 2UR; telephone Bracknell 20242.

A list of observing stations reporting hourly totals of solar radiation as supplied by the Meteorological Office, is set out in the following table with the following abbreviations:

T.   Total (global) solar radiation on a horizontal surface.

D.   Diffuse solar radiation on a horizontal surface. (Global radiation with the direct component from the sun removed by a shade-ring.)

L.   Total illumination on a horizontal surface. (Measured by an illuminometer with a spectral response similar to a human eye.)

F.   Diffuse illumination on a horizontal surface.

B.   Radiation balance. (Incoming minus outgoing radiation of all wavelengths.)

N.S.E.W. Global solar radiation on vertical surfaces facing North, South, East and West respectively.

SS.   Duration of bright sunshine in hours, measured by a Campbell-Stokes sunshine recorder.

239

Values of T., D., B. and N.S.E.W. are expressed in milliwatt–hours per square centimetre, while L. and F. are expressed in kilolux–hours.

The time standard used throughout is Local Apparent Time (L.A.T.).

UK STATIONS REPORTING HOURLY TOTALS OF SOLAR RADIATION
(EXCLUDING OCEAN WEATHER SHIPS)

Station	Latitude	Longitude	Elevation above sea level (metres)	Elements measured	Date of first observation
Lerwick	60°08' N	01°11' W	82	T.D.L. B.SS. (L. - 1.1.58, B. - 1.1.64)	Jan. 1952
Aberdeen	57°10' N	02°05' W	35	T.	June 1967
Dunstaffnage	56°28' N	05°26' W	3	T.	Apr. 1970
Dundee (Mylnefield)	56°27' N	03°04' W	30	T. B.	July 1973
Eskdalemuir	55°19' N	03°12' W	242	T.D.L. B.SS. (L. - May 58, B. - Feb. 64)	Jan. 1956*
Aldergrove	54°38' N	06°13' W	71	T.D.L. B.SS. (L. - June 71)	Jan. 1969
Cambridge	52°13' N	00°06' W	23	T.D. SS.	Jan. 1952 - Dec. 1971
Aberporth	52°08' N	04°34' W	115	T.D. SS.	July 1957θ
Cardington	52°06' N	00°25' W	29	T.D. SS.	Jan. 1972
Hurley	51°32' N	00°49' W	43	T.	Mar. 1969
London Weather Centre	51°31' N	00°07' W	77	T.D.L. SS. (L. - 1.1.67)	Jan. 1958φ
Kew	51°28' N	00°19' W	5	T.D.L.F.B.SS. (F. - Mar. 64)	Jan. 1947Δ
Bracknell	51°23' N	00°47' W	73	T.D.L.F. SS. (N.S.E.W. - Jan. 67)	Feb. 1965
Jersey	49°12' N	02°11' W	85	T.D.L. B.SS. (L. - Jan. 69)	Jan. 1968

*Data from June 1952 in manuscript
θData from Jan. 1953 in manuscript : nil Mar. - Dec. 1958
φData from Jan. 1950 in manuscript
ΔData from July 1932 in manuscript

APPENDIX 3

SOME REFERENCE SOURCES

INFORMATION IN THE UK

(1) The UK section of the International Solar Energy Society (ISES) can be contacted at 19 Albemarle Street, London W1X 3HA. Telephone 01 499 6601. By January 1982 they had published 28 different sets of Conference Proceedings. Sets of Proceedings from several of the major International Solar Conferences can be obtained from them.

(2) A Solar Energy Information Office has been established by the Department of Industry to serve industry with data and advice at the Department of Mechanical Engineering and Energy Studies, University College, Cardiff CF1 1XL. Telephone (0222) 44211. The Solar Energy Unit at Cardiff University is the leading research group in the UK.

(3) The Solar Trade Association 26 Store Street, London WC1E 7BT. Telephone 01 636 4717. Promotes the understanding of the advantages of solar energy and can advise on manufacturers of solar equipment.

(4) The Building Research Establishment have a major solar energy research programme at Garston, Watford WD2 7JR and publish Current Papers on topics related to energy use in buildings. Telephone (09273) 74040.

(5) Copper - advice on properties and on the availability of copper based solar heating panels and systems from The Copper Development Association, Orchard House, Mutton Lane, Potters Bar, Herts EN6 3AP. Telephone Potters Bar 50711.

(6) Airwrap packing sheet - a UK supplier is Abbotts Packaging Ltd, Gordon House, Oakleigh Road South, London N11. Telephone 01 368 1266. (Smallest bubble grade C120.) 1981 cost 37 p/m^2. Difficult to purchase in small quantities.

(7) Butyl sheet - a UK supplier is Butyl Products Ltd, 11 Radford Crescent, Billericay, Essex. Telephone Billericay 53281. 1981 cost approximately £2.00/m^2.

Appendix 3

OTHER INFORMATION SOURCES

(1) The International Referral System
The system is a world-wide network comprising a central United National
Environment Programme (UNEP) headquarters and international, national,
regional and sectoral focal points. The address of the nearest regional
office can be obtained from: The Director, UNEP/INFOTERRA, United
National Environment Programme, Box 30552, Nairobi, Kenya. Telephone
333930. Telex 22068/22173. Cable Uniterra Naiport.

(2) World Meteorological Organization
Geneva, Switzerland. Gives monthly averages for global insolation.

(3) The Solar Energy Industry Association
1001 Connecticut Avenue, N.W. Suite 632, Washington, DC 20036.

Similar to UK Trade Association in its activities.

(4) National Solar Heating and Cooling Information Center
P.O. Box 1607, Rockville, MD 20850.

The United States Government supports this very comprehensive solar
information section.

(5) International Solar Energy Society
P.O. Box 26, Highett, Victoria 3190, Australia.

ISES advises on the addresses of the nearest regional sections. These
sections provide a useful forum for the exchange of experience. Their
journals, particularly *Solar Energy* contain useful design information.

(6) Brace Research Institute   and   (7) Renewable Energy Resources Center
McDonald College of McGill University   Asian Institute of Technology
Ste. Anne de Bellevue, Quebec,   P.O. Box 2754, Bangkok,
Canada H9X 3M1.   Thailand

These two institutions have particularly good experience of working with
developing countries.

The Brace Research Institute has published many practical papers on
solar applications and the Renewable Energy Resources Center acts as an
information centre for the tropical countries in Asia and the Pacific,
as well as similar tropical regions in Africa, Central and South
America.

SOME EARLY NATIONAL REPORTS

(1) Solar Energy: a UK assessment
UK Section of ISES. First published in 1976 and extensively revised
by Leslie F. Jesch as *Solar Energy Today* in 1981.

(2) Description of the solar energy R & D programs in many nations
F. de Winter and J. W. De Winter, eds. Available from National
Technical Information Service, US Department of Commerce, 5285 Port
Royal Road, Springfield, VA 22161 (February 1976).

(3) Solar Energy for Ireland
Report to the National Science Council by Eamon Lalor, Government
Publications Sale Office, GPO Arcade, Dublin 1 (February 1975).

(4) Report of the Committee on Solar Energy Research in Australia
     Australian Academy of Science Report No. 17 (September 1973).

(5) Solar Energy as a National Energy Resource
     NSF/NASA Solar Energy Panel Report. Available from National Technical
     Information Service, U.S. Department of Commerce, 5285 Port Royal Road,
     Springfield, VA 22161, Report PB 221659 (December 1972).

                                    JOURNALS

(1) Solar Energy
     The Journal of Solar Energy Science and Technology, published bimonthly
     up to 1978 and monthly thereafter by Pergamon Press for the International
     Solar Energy Society.

(2) Sunworld
     The bimonthly news magazine of the International Solar Energy Society,
     published by Pergamon Press.

(3) Sun at Work in Britain
     The magazine of the UK section of ISES, 19 Albemarle Street, London
     WIX 3HA, published by Pergamon Press.

(4) Gelioteckhnika
     The Russian journal published in English translation by Allerton Press,
     New York.

(5) Bulletin of the Cooperation Mediterraneanne pour l'Energie Solaire
     Published by COMPLES, 32 Cours Pierre-Puget, 13006 Marseilles, France.

(6) Energy
     A multidisciplinary International Journal published by Pergamon Press.
     The January 1982 issue, Vol. 7, No. 1, is a special issue concentrating
     on the solar energy/utility interface.

                                    BOOKS

(1) Direct Use of the Sun's Energy
     Farrington Daniels. Yale University Press (1964) and reprinted
     Ballantine Books, New York (1974). A widely acclaimed basic source
     book, particularly interesting now for the historical viewpoint.

(2) Solar Energy
     H. Messel and S. T. Butler, eds. Shakespeare Head Press, Sydney (1974)
     and Pergamon Press (1976). A broad introduction covering developments
     up to the early 1970s.

(3) The Food and Heat Producing Solar Greenhouse (1976)
     Bill Yanda and Rick Fisher, John Muir Publications, PO Box 613, Sante
     Fe, NM 87501. Both attached and free-standing greenhouses are
     described, with many clear diagrams and photographs illustrating
     practical methods of construction.

(4) Engineer's Guide to Solar Energy (1979)
Y. Howell and Justin A. Bereny, Solar Energy Information Services,
PO Box 204, San Mateo, CA 94401. A very comprehensive useful guide to
all aspects of the practical applications of solar energy. It includes
an extensive annotated bibliography, world solar radiation and climatic
data, a product directory of United States companies and applications
of passive heating and cooling of buildings.

(5) Practical Solar Heating (1979)
Kevin McCartney with Brian Ford, Prism Press, Stable Court,
Chalmington, Dorchester, Dorset, UK. An excellent practical handbook
for designing and constructing a solar water heating system.

(6) Biological Energy Resources (1979)
Malcolm Slesser and Chris Lewis, E. & F. Spon 1979. An introduction
in depth to solar biological applications.

(7) Solar Energy (1979)
J. I. B. Wilson, Wykeham Science Series. An introduction in depth to
the photovoltaic applications.

(8) The Sun: our future energy source (1979)
David K. McDaniels, John Wiley, New York. Gives a "liberal arts"
student a broad background of the basic solar concepts.

(9) European Solar Radiation Atlas (Volume 1) (1979)
Commission of the European Communities, Ed. S. Palz. This is the first
volume in a proposed series and gives the global radiation on
horizontal surfaces during the year.

(10) The Passive Solar Energy Book (1979)
Edward Mazria, Rodale Press. An invaluable guide to passive solar
architecture, with a wealth of practical information clearly
illustrated.

(11) A Golden Thread (1980)
Ken Butti and John Perlin, Cheshire Books and Van Nostrand Reinhold.
Particularly informative on the early historical applications, this is
a scholarly reference text presented in an entertaining style.

(12) Solar Engineering of Thermal Processes (1981)
J. A. Duffie and W. A. Beckmann, John Wiley, New York. The leading
basic reference source for both undergraduate and postgraduate courses.

(13) Solar Energy (1981)
D. Rapp, Prentice-Hall. Another basic reference, suitable for both
undergraduate and postgraduate courses with an emphasis on computing
techniques for thermal applications.

GLOSSARY

Absorptivity
The ratio of the amount of radiation absorbed by a body to the radiation
incident upon it. (absorptivity = 1 - reflectivity.)

Air mass
The ratio of the actual distance traversed through the Earth's atmosphere
by the direct solar beam to the depth of the Earth's atmosphere, normal to
the surface.

Albedo
The ratio of the radiation reflected from the Earth to the total amount of
radiation incident upon it.

Algae
Filamentous or unicellular water plants, normally fast growing.

Altitude
The angle between a straight line from the sun to the centre of the earth
and the tangent to the surface of the earth (90° - zenith).

Anaerobic fermentation
Fermentation caused by micro-organisms (bacteria) in the absence of oxygen.

Aphelion
The point in the orbit of the Earth where it is at its greatest distance
from the sun.

Array
A bank or set of solar modules or panels.

Attenuation
The reduction of radiation flux over a given path length, due to absorption
and scattering.

Autarchic
Self-sufficient. Applied to a house, it would be independent of all mains
services.

245

Azimuth angle
The angle between the south meridian, measured in a horizontal plane west-
wards, and the direction of the sun (note that this convention is normally
used in solar work).

Bioconversion
The conversion of solar energy into chemically stored energy through
biological processes. Various fuels and materials can be produced by
bioconversion.

Black body
A term denoting an ideal body which would absorb all and reflect none of the
radiation falling upon it. Its reflectivity would be zero and its
absorptivity would be 100%. An alternative definition is a body which at
any one temperature emits the maximum possible amount of radiation, i.e. its
emissivity is 1.0. The total emission of radiant energy from a black body
takes place at a rate expressed by the Stefan-Boltzmann Law.

Celestial Sphere
A sphere of infinite radius with its centre located at a point within the
solar system. The reference frame for all systems of astronomical spherical
coordinates are based on the celestial sphere.

Collector
A solar collector or absorber is used to collect solar radiation. In the
process, the radiation undergoes a change in its energy spectrum.

Concentration ratio
The ratio of the irradiance at the focus of the concentrator to the direct
radiation received at normal incidence on the surface.

Concentrator
A device for focussing solar radiation.

Cuprosolvent
Tends to dissolve copper.

Declination
Solar declination is the angular distance between the sun and the plane of
the celestial equator.

Diffuse radiation or insolation
Solar radiation which arrives on Earth as a result of the scattering of
direct solar radiation by water vapour and other particles in various layers
of the atmosphere. It is also known as indirect radiation or sky radiation.

Direct radiation or insolation
Radiation from the sun which eventually reaches the Earth's surface without
being scattered. This is also known as direct beam radiation.

Emissivity
The ratio of the radiation emitted by a body to that which would be emitted
by an ideal black body the same temperature and under identical conditions.

Evacuated tubular collector
A collector which has two outer concentric glass tubes. A third central
tube contains the heat transfer fluid, while the evacuated space between the
outer two tubes reduces heat losses. Characteristically, their efficiency
is in the order of 50% with temperature differences between the heated fluid
and the surroundings greater than 100°C.

Flat plate collector
Any non-focusing flat surfaced solar collector.

Global radiation
The sum of the intensities of the vertical component of direct solar
radiation and the diffuse radiation. An alternative definition is radiation
at the Earth's surface from both sun and sky.

Greenhouse effect
An expression given for the solar heating of bodies shielded by glass or
other transparent materials which transmit solar radiation but absorb the
greater part of the radiation emitted by the bodies.

Heat pipe
A device for transferring heat by means of the evaporation and condensation
of a fluid in a sealed system.

Heat pump
A reversed heat engine; it transfers heat from a lower temperature to a
higher temperature by the addition of work. The amount of heat delivered at
the higher temperature divided by the work input is called the coefficient of
performance.

Heliostat
A mobile array of mirrors used to reflect a beam of sunlight in a fixed
direction as the sun moves across the sky.

Incidence angle
Angle between the perpendicular to the surface and the direction of the sun
at that instant.

Infra-red radiation
The band of electromagnetic wavelengths lying between the extreme of the
visible region (circa 0.76 μm) and the shortest microwaves (circa 1000 μm).

Insolation
Originally defined as one of the processes of weathering, it is now generally
regarded as another term for solar radiation, including the ultra-violet and
visible infra-red radiation regions. Total insolation is a term sometimes
used instead of global radiation.

Irradiance
Radiant energy passing through unit area per unit of time.

Irradiation
The process of exposing to radiation.

Perihelion
The point in the earth's orbit when the earth is closest to the sun.

Photobiology
A biological subject covering the relationship between solar radiation,
mainly in visible wavelengths, and biological systems.

Photochemistry
A chemical subject dealing with chemical reactions induced by solar radiation.

Photosynthesis
The conversion of solar energy by various forms of plant and algae into
organic material (fixed energy).

Photovoltaic cell
Also known as photocell or solar cell. A semiconductor device which can
convert radiation directly into electromotive force. An alternative
definition is a device used for detection and/or measurement of radiant
energy by the generation of an electrical potential.

Power tower or solar tower
A tall tower, perhaps 500 m high, positioned to collect reflected direct
solar radiation from an array of heliostats. The top of the tower contains
the heat exchange chamber and the hot working fluid is used in a conventional
electrical generating system at ground level.

Pyranometer
An instrument used for measuring global radiation, also known as a
solarimeter.

Pyrheliometer
An instrument used to measure the direct irradiance of the sun along a
surface perpendicular to the solar beam. Diffuse radiation is excluded from
the measurement.

'R' factor
R is the thermal resistance expressed as the temperature difference required
for 1 watt to pass through 1 square metre per hour.

Radiant energy
Energy transmitted as electromagnetic radiation.

Radiation
Radiant energy.

Reflectivity
The ratio of the radiation reflected from a body to the radiation incident
upon it. (reflectivity = 1 - absorptivity.)

Retrofitting
Erecting a solar collecting system on to existing buildings.

Scattering
Interaction of radiation with matter where the direction is changed but the
total energy and wavelength remain unaltered. An alternative definition is
the attenuation of radiation other than by absorption.

Selective surface
A surface which has a high absorptivity for incident solar radiation but also
has a low emissivity in the infra-red region.

Semiconductor
An electronic conductor whose resistivity lies in the range 10^{-2} to 10^9 ohm-cm (between metals and insulators).

Solar constant
The amount of solar radiation which is received immediately external to the Earth's atmosphere and incident upon a surface normal to the radiation taken at the mean Earth-sun distance. It is not a true constant as it varies, mainly due to sun-spot activity. A mean value of 1.373 kw/m^2 ± 2% is normally taken, although values outside these limits can occasionally occur.

Solar furnace
A device used for achieving very high temperatures by the concentration of direct radiation.

Solarimeter
Another name for Pyranometer.

Solar pond
An artificially enclosed body of water containing a stratified salt solution. Solar energy can be stored as heat in the pond, as the stratified salt solution reduces heat losses.

Spectrometer
An instrument for the measuring of radiation intensity over small wavelength intervals.

Spectroradiometer
An instrument for measuring spectral irradiance.

Thermosyphon
Natural liquid circulation caused by the small difference in density between a hot and a cold liquid. In a solar collector thermosyphon system, the collector is placed below the water storage tank and the solar heated water rises to the top of the tank, displacing colder water from the bottom of the tank to the bottom of the collector.

Turbid atmosphere
An atmosphere containing particulate material (aerosols).

Turbidity
A factor describing the attenuation of direct radiation due to suspended particulate material. Sometimes defined as the factor by which the air mass is multiplied to allow for attenuation in the turbid atmosphere.

'U' value
The reciprocal of the 'R' factor.

Ultra-violet radiation
The band of electromagnetic wavelengths lying next to the visible violet (0.10-0.38 μm).

Visible region
The range between the ultra-violet and infra-red regions. This region can also be defined as that which affects the optic nerves (0.38-0.76 μm).

Zenith angle
The angle between a line from the sun to the centre of the Earth and the normal to the surface of the Earth (90° - altitude).

LIST OF SYMBOLS

Most of the symbols used in this book are given below. Other symbols, including various constants, are defined in the text.

A area

C capital cost

D diffuse solar radiation, diameter

E internal energy

e emmitance

F collector heat removal factor, annual savings

F_c cost of competitive energy

f annual inflation rate

G global solar radiation

G_c incident solar radiation normal to collector

G_i total annual incident solar radiation normal to collector

H average hours in the year

I direct solar radiation

i net effective interest rate

ℓ loss of efficiency factor

M capital repayment factor

N mean length of day

n mean daily bright sunshine hours, a period of years

Q useful heat collected per unit area, heat transferred

r annual interest rate

T temperature, maintenance cost

U overall heat loss coefficient

V velocity

W external work

Y constant annual payment

Z original annual payment

z zenith distance

Greek

α absorptance

γ solar altitude

η efficiency as defined in the text

ρ reflectance, density

τ transmittance

Subscripts

1   initial, source

2   final, sink

a   ambient

i   inlet

m   mean, per m^2

R   rated

S   starting

INDEX